Challenges in Fluid Dynamics

R.Kh. Zeytounian

# Challenges in Fluid Dynamics

## A New Approach

Springer

R.Kh. Zeytounian
Université de Lille I
LML – cité Scientifique
Villeneuve d'Ascq, France

ISBN 978-3-319-31618-5 ISBN 978-3-319-31619-2 (eBook)
DOI 10.1007/978-3-319-31619-2

Library of Congress Control Number: 2017962965



Printed on acid-free paper

This Springer imprint is published by Springer Nature
The registered company is Springer International Publishing AG
The registered company address is: Gewerbestrasse 11, 6330 Cham, Switzerland

*This book is dedicated to the memory of: Arto Mkhitarian S. N. Mergelyan and I. A. Kibel and also: my friend and colleague Jean-Pierre Guiraud*

# Preface

> Zeytounian's (2012) book makes a good case that his RAM approach deserves more serious computational attention than it has received. His presentation is generally clear, though Springer should have devoted more effort to cleaning up the English and the typography.
>
> Robert E. O'Malley, Jr—Books Reviews of SIAM Review 54(3) page 613
>
> The most outstanding Chap. 4 of Zeytounian's (2014) book develops the basic Navier-Stokes–Fourier (NSF) equations in a historical context. Overall, readers will realize the unique value of Zeytounian's work and perspective and will come to appreciate his willingness to tackle very difficult problems, aiming to help Fluid Dynamical Numericians. An example is modeling turbo-machinery using the number of blades on a rotor as a large parameter.
>
> Robert E. O'Malley, Jr—Books Reviews of SIAM Review 56(3) page 569

My last two books were published by Springer-Verlag in Heidelberg in *2012* and *2014*. As a *research monograph, the 2012 book* is mainly devoted to the Navier-Stokes-Fourier (NSF) system of equations for a Newtonian fluid in the case of compressible, viscous, and heat-conducting flows. The main tool developed there was what I call the *Rational Asymptotic Modelling (RAM)* approach. In this 2012 book, the main objective has three facets. *First*, to "*deconstruct*" the NSF system of equations in order to bring some unity to the various partial, simplified, and approximate models used in classical Newtonian fluid dynamics; this first facet of the main objective is obviously quite a challenge and has very important pedagogical consequences for university education. The *second facet* of the main objective is to outline a consistent rational asymptotic theory for modeling fluid flows on the basis of a typical NSF initial and boundary value problem. The *third facet* is an illustration of our *rational asymptotic modeling (RAM) approach* for various tough technological and geophysical problems from aerodynamics, thermal and thermocapillary convections, and also meteo-fluid dynamics.

Concerning *the 2014 book*—a "*Scientific Autobiography*"—the reader is guided through my somewhat *unconventional career*: first *discovering* fluid mechanics, and then devoting more than *fifty years* to *intense work in the field*. Using both personal and general historical contexts, this account would be of benefit to anyone interested in the early and contemporary development of an important branch of theoretical and computational fluid mechanics.

*The present book (written during the years 2013–2016), together with the two mentioned above, is a major part of a "trilogy" providing a challenging modern point of view of fluid dynamics that examines the origins of working models and their classification*. This lecture course seems to correspond well to a suggestion of *Paul Germain* (see the conclusion in his last paper [1]) concerning "*fluid mechanics inspired by asymptotics*" as a new way to look at problems, which has proven itself very useful in g*as dynamics, hyposonic, supersonic,* and *hypersonic aerodynamics, magnetofluid dynamics, progressive waves*, *internal structure* of *shock waves, hydrodynamic instability, combustion with high activation energy, waves near breaking, atmospheric motions, convection in liquids, nonlinear acoustics, and so on!* More precisely then, in the present book *Challenges in Fluid Dynamics*—a close companion to the *2012 book*—the reader will find m*any new facets* and *applications* of our *rational asymptotic modelling* (RAM) approach.

In particular, my purpose has also been to provide a *logically consistent answer* to two questions mentioned below concerning the curious "*fragmentation*" of *Newtonian fluid dynamics,* which generates a *never-ending and varied list of fluid dynamics problems describing certain very particular models of fluid flows!* These *two very reasonable questions* were *raised in a book written over 40 years ago* by *Martin Shinbrot* [4]:

> Where is he to begin [concerning, the never-ending list of particular fluid flow models], and which of these is fundamental, and which created to aid in the explication of some specific physical problem?

In reality, it seems to me that today *most fluid dynamicists and applied mathematicians only learn the answers to these two questions after they have worked on a specific fluid dynamics problem presented to them on an ad hoc basis. The first step here is* the *dimensionless analysis of NSF problems with specific non-dimensional parameters relating to the various physical effects and associated with the corresponding terms in NSF problems. The second step involves* the *emergence from these dimensionless NSF problems of a large family of approximate rational reduced working models, which leads to a classification according to limiting values (mainly zero or infinity) of a large number of dimensionless reduced parameters (in NSF problems)*: *Reynolds* number (viscosity), *Mach* number (compressibility), *Prandtl* number (heat conduction), *Strouhal* number (unsteadiness), *Froude/Boussinesq* numbers (gravity), *Rossby/Kibel* numbers (earth rotation), *Grashof/Rayleigh* numbers (thermal convection in liquids), *Marangoni* number (thermocapillary effect), and so on.

In this way, we also get a *clear view of* the *origins* of the *emerging reduced working models, and this gives another understanding of the NSF equations, and undoubtedly a new status to the Stokes "fluidity" concept, discussed in Sect. 1.3, as a basis for classical Newtonian fluid mechanics*. I think, on the one hand, that these emerging approximate rational reduced working models, derived using the RAM approach, should also be very welcome for *"pure" mathematicians*, when they discover that there are rigorous results concerning the existence, regularity, and uniqueness of solutions of various initial-boundary value problems in fluid dynamics! This seems to me obvious if we take into account the fact that, in so-called *"mathematical fluid dynamics"*, the considered fluid flow problems are very often in no way *rigorously* linked with the fundamental classical NSF equations for Newtonian fluid flows!

At the same time, the approximate rational and consistent reduced working models we derive from very stiff NSF problems should also be valuable for *numericians* dealing with *numerical simulations/computations* using high-memory *high-speed computers.*

In a certain sense, Chap. 4 of our 2014 book, devoted to the *interrelationship between NSF equations and our RAM approach*, can be considered as a short preliminary presentation of the *NSF system of equations via the RAM approach,* in the space of just 54 pages. And in Sect 3.4, the reader will find a short, but nevertheless illuminating account of the RAM approach, through a rather *laborious* application to the derivation and justification of some approximate reduced working meteo-atmospheric models for the weather forecast.

*Obviously, a Zeytounian "trilogy" is basically very different from a traditional course of lectures on fluid dynamics, and gives a different, rather challenging view of fluid dynamics, in which the NSF system of equations is given the most important place!* Indeed, the *main novelty in* our approach comes chiefly from the fact that *the fundamental system of NSF equations governing the Newtonian motion of a compressible, viscous, and heat conducting fluid flow* is *given a central role in our discussion. This leads to a new vision of the universality of NSF equations as a fundamental mathematical system for the study of many aspects of Newtonian fluid dynamics, but also to its "deconstruction" via the RAM approach!*

However, a *strong restriction* is assumed (at least at the present time!), namely, that the considered fluid flows are *laminar!* This is unfortunate because some fluid flows appear to be very *tumultuous, changing rapidly from point to point* or *from time to time* at a given position in 3D time-space! Such *non-laminar* and *very tumultuous fluid flows*, so-called *turbulent fluid flows* (where the usual laminar motion becomes turbulent if for instance the *Reynolds* numbers is *too high* and there is a laminar portion upstream, followed by a *transition to turbulence*) *definitely fall outside the scope of our considerations in the present book.* Discussion of turbulent motion is necessarily a very *long business*, but the reader can find various valuable books dealing with turbulence, despite the fact that turbulence is a huge subject of intensive ongoing research. It seems to me that, in the book by *Sagaut et al.* [3], the reader can find a comprehensive description of modern strategies for turbulent flow simulation, ranging from turbulence modelling to the most advanced numerical methods.

As a final remark we observe that a consequence of our macroscopic-continuum approach, equations of state and transport coefficients (such as viscosity and heat conduction) are given as phenomenological or experimental data and are not related to microscopic equations, i.e., laws governing molecular interactions.

In *Golse* [2], a paper with 142 pages, the curious and motivated reader will find a good account of the *Boltzmann equation and its hydrodynamic limit* However, *Golse* mentions (on p. 238 of [2]) *rather disappointedly* that: "*The derivation of the NSF system of equations from the Boltzmann equation leads to dissipation terms that are of the order of the Knudsen number (Kn≪1; ratio of Mach/Reynolds), and therefore vanish in the hydrodynamic limit, Kn→0. In other words, the NSF system of equations is an asymptotic expansion of the Boltzmann equation in the Knudsen number, and not a limit thereof!*"

I stress again that the deconstruction carried out in the RAM approach is based on a *rationally argued, consistent, non-ad-hoc, and non-contradictory approach to the full unsteady NSF system of equations for modelling Newtonian fluid flows*, tackled in the spirit of *asymptotics. Our main goal is the modelling, and not the search for solutions to the derived reduced working model equations with initial and boundary conditions. The aim is thus to assist numericians in the context of their simulation/computations using super-high-power computers!*

It seems to me that the RAM approach opens up new vistas for the derivation of many valuable working model problems. Such models offer a large panel of tools for fluid dynamicists interested in engineering, environmental, meteorological and atmospheric, or geophysical applications. It gives theoretically oriented researchers a way to exchange with numericians expert in high-speed computing, in order to provide them with models that are much less stiff than "*brute force*", starting directly from the full dimensionless unsteady NSF system of equations and *boundary* conditions!

Let us end this preface with a brief description of the *nine chapters* in the present book, following the *introduction. Chapter 1* describes some *historical steps* relating to the discovery of the NSF equations, i.e., from *Newton (1686)* to *Stokes (1845)*, via *Euler (1755)*, *Navier (1821)*, *Cauchy (1828)*, *Fourier (1833)*, and *Saint-Venant (1843)*. I hope that this chapter will stimulate the interest of many readers, and why not also motivate some young mathematicians, to carry out theoretical investigations of more realistic applied problems involving Newtonian fluid flows governed by the NSF equations!

In *Chap. 2*, the reader can find first a formulation of a typical initial-boundary value, unsteady NSF problem in the case of a *thermally perfect gas*, and then the NSF equations applicable to an *expansible liquid*, Navier-Stokes (NS, *physically unreal, but* often used by mathematicians) *barotropic* compressible fluid flow, and NSF equations applied in *nonlinear acoustics as a branch of fluid dynamics*.

In *Chap. 3*, we briefly present the foundations of the RAM approach to emphasize our main *postulate* and *some key steps in its realization,* relating principally to the *dimensionless approach*, *asymptotics,* and various *limiting processes* relative to some of the most significant reduced *dimensionless parameters* in the NSF equations and boundary conditions. The case of dimensionless NSF equations for *atmospheric motions* (Sect. 3.4) is carefully considered to illustrate the possibilities of the RAM approach.

In *Chap.* 5, I have tentatively sketched a "*theory of models*" emerging from the full unsteady dimensionless NSF problem in the RAM approach. In Figs. 5.1 and 5.2, the reader will find many valuable approximate working models, which give a *preliminary idea* about the *origins and classification* of a family of models arising as a RAM limit of the dimensionless NSF problem. In particular, Fig. 5.2 shows the approximate reduced working models for high and low Reynolds numbers (Re $\gg$ 1 and Re $\ll$1), and also for low Mach number (M$\ll$1).

Chapter *6*, is devoted to the RAM "deconstruction/modelling" of *four typical*, but very stiff NSF problems, namely, Re $\gg$1 and M $\ll$ 1, meteo-fluid dynamics models, and models for a weakly expansible liquid heated from below. The approximate working models derived for the above-mentioned typical cases are sketched in Figs. 6.1 to 6.4.

*Chapter 8* is a *miscellany* of major applications of the RAM approach I have been involved in over the past 50 years (1960–2010) (see *Sect. 6.1*). Chapters *4*, *7* and *9* are devoted to some concluding remarks about Chaps. *2*, *3*, 5, *6*, and *8*.

*I have written the present book as the last,* coming at the end of 40 years of writing, from 1974 to 2014, which has resulted in 14 books, mainly devoted to rational modelling of Newtonian fluid flows. I hope that the above-mentioned "trilogy" will be useful to students in the last years of university and young researchers working in the field of fluid dynamics and motivated by modelling, numerical analysis, and simulation/computation of very stiff complex real technological and geophysical flows. Our "trilogy" should also prove useful as a reference for scientists and engineers working in the various fields of fluid dynamics.

I owe thanks first to Dr. Christian Caron, executive publishing editor for physics at Springer, for his continued support during many years in the edition of my various books, and who suggested to Dr. Jan-Philip Schmidt, Springer editor for interdisciplinary and applied sciences, the publication of the present "Challenges" in the applied mathematics and engineering program. As for my previous books, I had several fruitful discussions with *J. P. Guiraud*, whose help was decisive in rendering my own ideas more precise!

My sincere appreciation goes to Dr. Jan-Philip Schmidt, Springer DE, and his copyediting and production team for their professional and kind assistance.

Paris/Yport R.Kh. Zeytounian
October 2015

## References Cited in the Preface

[1] Germain P (2000) The "New" mechanics of fluids of Ludwig Prantdl. In: Meier G (ed) Ludwig Prantdl ein Führer in der Stromungslehre. Vieweg, Braunschweig, pp 31–40
[2] Golse F (2005) The Boltzmann equation and its hydrodynamic limits. In: Dafermos CM, Feireisl E (eds) Handbook of differential equations, evolutionary equations, vol. 2, chap. 3. Elsevier B.V, pp 159–301
[3] Sagaut P, Deck S, Terracol M (2006) Multiscale and multiresolution approaches in turbulence. Imperial College Press

[4] Shinbrot M (1973) Lectures on fluid mechanics. Gordon & Breach Science Publishers, New York
[5] Zeytounian RKh (2012) Navier-Stokes–Fourier equations—a rational asymptotic modelling point of view. Springer, Berlin
[6] Zeytounian RKh (2014) Five decades of tackling models for stiff fluid dynamics problems—a scientific autobiography. Springer, Berlin

# Contents

# Introduction

Classically, the three main *continuous macroscopic systems of equations of fluid dynamics*, namely, *Euler* (*non-viscous fluid*), *Navier* (*incompressible and viscous*), and *Navier-Stokes-Fourier* (*NSF,* not only *viscous* and *compressible,* but also *heat conducting*), are first established by applying *Newton's second law of motion.* In his major treatise, *Philosophiae Naturalis Principia Mathematica, published in 1687* [1], *Newton* gave a mathematical description of the laws of mechanics and gravitation, and applied this theory to explain *planetary and lunar motion.* For most purposes, *Newtonian classical mechanics* has survived up to now, despite the advent of *Einstein's revolutionary relativity theory* and *Dirac's famous quantum mechanics* at the beginning of the twentieth century.

Indeed, Newtonian mechanics is still a very good approximation for the mathematical description of *terrestrial* and *cosmological* phenomena, and the reader will find in *Serrin* [2] a clear presentation of the NSF system of evolution/unsteady equations, which is a complete closed system *for the density* $\rho$ *(the continuity equation), the velocity* $\boldsymbol{u}$ *(the equation of motion), and the temperature* $T$ *(the energy balance equation).* Unfortunately, this NSF problem is, in fact, a *strange puzzle* that can only be solved and understood by hard thought! A *challenging* approach is to deconstruct the NSF problem using the RAM approach, in order *to extract approximate rational non-ad hoc working models. This also leads to a classification of many useful fluid flow problems and reveals their origins.*

The NSF problem, formulated in Chap. 2, will be the main focus of attention, using a systematic dimensionless analysis, asymptotics, limiting processes in terms of various dimensionless parameters, and similarity relations. This book is written by a fluid dynamicist mainly from the point of view of useful applied mathematics, and we will obviously not be interested in abstract mathematical rigor. Instead, our RAM/deconstruction approach is rather a typical mathematical physics investigation, i.e., a rationally argued, non-ad hoc, and self-consistent modeling theory of the full unsteady NSF system of equations governing Newtonian fluid flows, carried out in the spirit of asymptotics, using the RAM approach!

Our *main objective is to give a clear statement of the underlying ideas,* using all the means at our disposal in the *framework of our imaginative rational asymptotic modelling (RAM) approach, coupled with a deconstruction of the NSF system of equations.* Clearly, the "trilogy" à la Zeytounian mentioned in the preface as a challenging fluid dynamics course is *basically very different* from a traditional course of lectures on fluid dynamics and gives *another view* of *theoretical fluid dynamics.* The main *originality* of our approach lies chiefly in the following fact: *the fundamental system of Navier-Stokes-Fourier (NSF) equations,* which governs the Newtonian compressible, viscous, and heat conducting fluid flow, remains at the *centre of focus. This unusual strategy means that we can deconstruct the NSF system of equations via the RAM approach and thus construct a logical scheme.*

*First, there is a classification of useful reduced approximate working model problems, in what seems like a never-ending list, describing a significant portion of traditional fluid dynamics. This deconstruction strategy also reveals a second striking feature intrinsic to the NSF problem, which clears up the problem of the origins of many useful reduced approximate working model equations. As a consequence, the deeper meaning of the NSF problem no longer remains a puzzle!*

This significant advance in our understanding of the role of the NSF equations should obviously have a very important pedagogical impact on the university teaching of Newtonian fluid dynamics. Naturally, we may start at the outset with the elaboration of the NSF system of equations (see for instance Chap. 1) which leads to a logical reasoned and mathematically well formulated university fluid dynamics course! More precisely, it will be shown that the above-mentioned never-ending list of reduced, simplified working models (according to Marvin Shinbrot's 1973 book, cited in the preface) is closely connected with the various limiting processes associated with the reduced, dimensionless reference parameters/numbers, in the framework of various dimensionless initial-boundary value NSF problems with the following parameters: Reynolds (viscosity), Mach (compressibility), Prandtl (heat conduction), Froude/Boussinesq (gravity), Strouhal (unsteadiness), Grashof/Rayleigh (dilatable liquids), and Rossby/Kibel (atmospheric motions), to mention but a few! The last case, concerning the modelling of atmospheric motions, is a very representative one, and is thus discussed in Chaps. 3 and 6.

Our approach also provides an effective tool for researchers embarked on the numerical modelling of very difficult stiff fluid flow problems. As an example, I mention here two rather complicated cases. The first is the theory of axial flow through a turbine using, as a large parameter, the number of turbine blades per rotor. The second tackles the question of what the weather will be like tomorrow, or for the next few days, in the context of weather forecasting! As will be shown in Sect. 3.4, the choice of a small parameter is rather laborious here!

For a deeper treatment of the fundamental ideas of our RAM approach, the reader is referred to our companion book published in 2012. As already, mentioned in the preface, *the present book, together with the two books published in 2012 and in 2014, is the third component of a "trilogy", viz., a modern course of lectures on fluid dynamics, devoted mainly to the origins of models emerging from NSF initial-boundary value problems and their classification.* Below, the reader will also find a consistent logical answer, using our *RAM approach*, to the *two main questions* raised by *Marvin Shinbrot* in 1973, which bring new light to Newtonian fluid dynamics and put the NSF problem back in its rightful—*central*—place!

It seems to me that interdisciplinarity, involving a close working relationship between numerical scientists and applied mathematicians, is a crucial issue in modern fluid dynamics. In this context, one of my main motivations in writing this final book in my "trilogy" is to present a n*ew way to look at the problems of Newtonian fluid dynamics using the deconstruction/RAM approach. My purpose is also to provide some reflections and opinions on the current situation in the rational theory of Newtonian fluid dynamics.*

*First of all, it is obvious that mathematical topics in fluid dynamics are currently very popular among those interested in pure mathematics and familiar with abstract nonlinear functional analysis, mainly based on an endless list of bounds, a priori estimates, classes of functions, norms, functional (weighted) spaces, and in particular inequalities, all of which are invaluable tools for mathematicians!* Concerning these *mathematical topics*, I fully agree with the pertinent remark in *Doering* and *Gibbon* [3, p. xii] that this topic "*has remained closed to the main-stream of applied mathematics and mathematical simulation/computation of various very difficult, stiff, real fluid flow problems*".

On the other hand, *numerical scientists* speak mainly about the significance of increasing the capacity of new *super-high-speed computers*, capable of carrying out *several million billion operations per second*. This is clearly a *very fruitful and significant advance for numerical fluid dynamics*, well expounded in *Birkhoff* [4], which gives a review (up to the beginning of the 1980s) of the present status and future prospects of numerical fluid dynamics, through a series of case studies. However, unfortunately, these numerical simulations, which need massive and costly computations, often use highly simplified approximate reduced working models, derived in an *ad hoc* way, even without realistic boundary conditions.

It is clear that, using such non-rational working models rather than consistent rational working models emerging from a full system of Navier-Stokes-Fourier (NSF) equations with initial and boundary conditions via the rational asymptotics modelling (RAM) approach, we may be suspicious of the real validity of the resulting numerically computed fluid flows. From the point of view of numerical science, it seems to me that little attempt is made to derive, on formally mathematical grounds, such simplified reduced rational working models.

In the end, while recognizing the need for such a computational endeavour, it seems that a greater effort should be devoted to the RAM approach. Instead of *ad hoc*, often worthless approximate reduced models, rational working models must necessarily be derived by neglecting terms which do not play a major role, at least to leading order, in the full dimensionless NSF equations and conditions. In fact, our RAM approach provides a way to take into account the neglected terms, in leading-order reduced working models, by deriving various *second-order working models* (see Chap. 8 for simple cases).

So why should one do that? We intend to give a rational answer to this crucial question with the help of our RAM approach, in order to carry out a deconstruction process of the NSF system of equations. Until this is done, and it is now possible using the RAM approach expounded in our *2012 book*, it will be difficult to convince the detached and possibly sceptical reader of the real value of calculated numerical results as an aid to understanding. It seems to me, maybe naively, that it is reasonable to ask the following simple question: *Is it satisfactory to spend so much money on computing with strongly reduced approximate models of the full unsteady NSF I-BV problem which are not built on a rational foundation?*

In this book, I wish to show that the RAM approach deserves more serious attention than it has yet received. We will be concerned (mainly in Chaps. 6 and 8) with modelling for simulations/computations of various Newtonian fluid flow phenomena relating to stiff technological, geophysical, thermal/thermocapillary convection, and acoustic problems, which are governed by the NSF equations or their two main reduced forms: Euler (non-viscous) and Navier (incompressible), associated with initial and boundary conditions.

The NSF equations govern not only compressible, viscous, and thermally perfect gases (aerodynamics and atmospheric motions), but also expansible liquids (as in the famous Bénard problem, which considers heating from below, of *Bénard, 1900/1901* [5]) and the problems of nonlinear acoustics (as a branch of fluid dynamics, judiciously mentioned by *Crighton* [6]). *The RAM approach to these problems constitutes a difficult task which poses a serious challenge, as is clearly apparent in our monograph FMIA 64*, 2002 [7], *and of course in Chaps. 6 and 8 of the present book*.

Rigorous pure mathematics, despite many valuable contributions, initiated by *Jean Leray* in two famous papers [8, 9] over eighty years ago, is not yet able to give a broad characterization of the behaviour of solutions involving stiff realistic aerodynamic conditions such as complicated flow *over an aircraft wing*, as in the case of the *French Rafale aircraft*!

Concerning numerical computations, we have to deal with such time-consuming operations that we have to rely on shortened approximate models, or even several of them associated with various localized space-time domains, reconstructing the global flow from these. But one may question the validity of such an approximate numerical simulation, as remarked above. In our *companion book* of *2012*, I opened a way out of this situation, after a 16 year collaboration with Guiraud, during the years *1970* to *1986*, publishing more than *26 papers* (see the bibliography of our joint paper *Guiraud* and *Zeytounian* [10] for references to my own and our joint papers).

My first attempt was a course in *two volumes* (written in French), *1986* [11] and *1987* [12], and the reader can find a short review in J Fluid Mechanics (Vol. 231, page 691, October 1991) with the following rather positive conclusion:

> The text is in French, but is easily understandable by anyone with elementary French. Equations are handwritten, but very clearly done. Some readers may feel that inadequate emphasis is placed on the physical interpretation of the results obtained. Nevertheless, in many of the areas covered in these two volumes there is a conspicuous lack of suitable expository material available elsewhere in the literature, and Zeytounian's notes are to be welcomed for filling these gaps, until a fuller and more specialized account appears in book form.

One may find in the published works and in teaching books—as in our own [13, 14]—a number of models amenable to solution, mainly with the aid of discrete numerical computation that is not too time-consuming. However, it is obvious that such working models emerge from the NSF problem by cancellation of some terms—in [13, 14] the reader can find numerous references concerning the theory and applications of non-viscous and viscous fluid flows (311 in [13] and 1156 in [14]).

A significant new approach was undoubtedly the introduction, mainly by *Kaplun* and *Lagerstrom, 1957* [15], of the mathematical technique of Matched Asymptotic Expansions (MAE) into the study of fluid dynamics problems. The magisterial and pioneering book by *Van Dyke* [16] contributed to spreading knowledge of that technique through the world of applied mathematicians and fluid dynamicists interested in fluid flow phenomena. At the time, the main purpose was not to support numerical simulation and computation, but rather to improve results by getting better solutions to many simplified problems which were rather poorly taken into account in the usual analytical approach! The fluid dynamical literature abounds with testimony to the success of this approach (see also, for instance, the edition of the *Van Dyke* book, *1975* [17], annotated by *Van Dyke* himself).

A typical example considered in *Van Dyke* [18], is the computation of heat transfer *in the nose* of a *space station during re-entry into the atmosphere*, a theme which was initiated and supported by the NASA Space Program. A solution is obtained from a leading order approximate model when the Reynolds number Re is very high. But the entry speed is so fast and the heating so great that, when the space shuttle encounters denser and denser atmosphere, one must take into account higher (at least second) order terms in the expansion with respect to the Reynolds number.

In *1962 numerical fluid dynamics was almost non-existent* due to the *lack of high-speed computers*! In the course of time, advances in computer technology have led to the development of increasingly accurate numerical solutions and thereby diminished the interest in approximate simplified closed analytical results. *In any case, for those interested in asymptotics, the development of increasingly accurate numerical solutions was not an argument for renouncing full asymptotics!*

As a personal example, I mention here a paper with *Guiraud* [19, 20], work carried out in ONERA (France) for the SNECMA, a French group, SAFRAN: "*Through-flow model in a turbomachine blade row*". This was arguably among the most complicated technologically important problems, and it was associated with a global one amenable to numerical computation. Another example is my *survey lecture, 1976* [21], devoted to *meteorology,* considered as a *discipline of fluid dynamics.* When writing this survey lecture, I wrote, with passionate belief:

> Asymptotics provide very powerful tools in the process of constructing mathematically consistent models for problems which are stiff from the point of view of numerical analysis and simulation. Whilst numerical and computational fluid dynamics seems now to be a mature discipline, for some time the growth in capabilities of numerical simulations should be associated with asymptotics in order to provide numericians with simplified, rational, and reliable models obtained via asymptotics.

This way of thinking as an aid to numerically oriented fluid dynamicists—via the books *1990* [22], *1994* [23], *2006* [24], *2009* [25], edited by Springer (Heidelberg)—arose from some simple observations.

To be more precise, when encountering a boundary layer, the numerician will obviously use *Prandtl's equations,* discovered in *1904* [26]. Another, far from obvious example occurs when dealing with numerical weather forecasting—prediction for the next few days—there is no need to complicate the computations by taking into account acoustic propagation! A way out of the difficulty was given as early as *1940* [27] by *Kibel*, extracting from the model a time related to the space scale and to the speed of sound. The efficiency as well as the accuracy of the trick was a consequence of *Kibel's* considerable mathematical skill.

Let us mention also the famous *Stokes paradox,* concerning flow behaviour far away from a body. According to *Kaplun* [28], this is solved by introducing both Stokes and Oseen behaviour associated with the vanishing Reynolds limit. Asymptotics related to a small parameter appear in all the examples advocated above, but neither *Kibel* nor *Prandtl* appealed to asymptotics, even as a quite simple argument, in contrast to *Van Dyke* and *Kaplun.*

One must recognize that mastering asymptotics necessitates a rather significant intellectual investment which might have to be made at the expense of the endeavour needed to master high speed computations. But ignoring asymptotics requires one to use an intuitive ad hoc approach, which *breaks down in certain localized regions of the space-time* domain occupied by the flowing fluid.

A point in favour of the RAM approach is that the whole flow domain is covered by several restricted sub-domains with a special mathematical working model, and thanks to matching, the numerical solution obtained for each sub-domain through its approximate model is an approximate leading-order solution to the one associated with the full global NSF problem in the relevant time-space region. In *Chap. 2*, such a typical initial-boundary value NSF problem is formulated in the case of a thermally perfect gas, as a working model (assumed "exact") for application of the *deconstruction/RAM approach.*

Indeed, such approximations and models emerging from various real NSF problems are generally constituted by the leading-order term of an asymptotic expansion, but even if one does not intend to improve the approximation, one has to deal with the problem of relating different approximations associated with different contiguous sub-domains, and in such a case, matching plays an important role, in the framework of the MAE technique.

When using high-speed computers to obtain the solution to the various models, it is not obvious how to join the numerical solutions associated with contiguous local ("inner") sub-domains with the main (so-called "outer") useful one! Of course, when dealing with Prandtl's boundary layer, one knows how to do this as a consequence of a long experience of such situations, but it is not clear what to do in more complex cases. Despite this, it seems to me possible to investigate this matching process numerically.

As noted in the preface, the present book is the third in a "*trilogy*", which is written in the same spirit, and is intended as a *modern course of lectures on fluid dynamics*.

The distance between "pure mathematicians", "applied mathematicians", and "research engineers" has currently grown much too large. However, I hope that both old and new investigators interested in Newtonian fluid dynamics model problems which emerge from the NSF system of equations via the RAM approach might learn much from it. In this context, it is well to understand the following: *The NSF system of equations is, as a matter of fact, at one and the same time, only a pretext, the centre and the pivot of the present "Challenges", and a step towards a new way to look at Newtonian fluid flow phenomena!*

Finally, I think that *the main originality of our view is the use of RAM associated with the deconstruction approach to bring out the perfectly illuminating relationship which exists between the NSF system of equations and useful approximate models, directly used and constituting a significant part of traditional fluid dynamics. The RAM/deconstruction approach clearly exposes the richness of the NSF system.*

*This leads naturally to a classification of the models in question, revealing their origins, and giving a deeper meaning to the NSF system which ceases to be a puzzle and acquires a central place in this new fluid dynamics!*

I should say that the various figures in Chaps. 2–8 are reprinted from the above-mentioned books in the references, and edited by Springer at Heidelberg during the years 1974–2014. On the other hand, concerning the *references* (in *square* brackets) in this introduction and in Chaps. 1–8, I have cited only the more significant books and papers! The curious and motivated reader can find in our two books [13] and [14] the references of other authors mentioned (see, for instance, the above reference: *Marvin Shinbrot* (1973), cited on p. 475 of [14]). Hence, the references cited at the end of Chap. 1 complete the list of references cited here in this introduction. The *references (*in *angle* brackets $< >$) for Part I, Part II, and Part III are listed at *the end* of Part III, before the *epilogue*.

## References

[1] Newton I (1687) Philosaphiae Naturalis Principia Mathematics. Joseph Streater, London
[2] Serrin J (1959) Mathematical principles of classical fluid mechanics. In: Handbuch der Physik, vol. VII/I, Springer, Berlin, pp 125–263
[3] Doering CR, Gibbon JD (1995) Applied analysis of the Navier-Stokes equations. Cambridge University Press, Cambridge
[4] Birkhoff G (1983) Numerical fluid dynamics. SIAM Rev 25(1), January
[5] Bénard H (1900/1901) Les tourbillons cellulaires dans une nappe liquide. Rev Générale Sci Pures Appl 11:1261–1271 and 1309–1328. See also: Les Tourbillons cellulaires dans une nappe liquide transportant de la chaleur par convection en régime permanent. Ann Chimie Phys 23:62–144
[6] Crighton DG (1981) Acoustics as a branch of fluid dynamics. J Fluid Mech 261–298

[7] Zeytounian RKh (2002) Asymptotic modelling of fluid flow phenomena. FMIA 64, Kluwer, Dordrecht
[8] Leray J (1983) J Math Pures Appl 12:1–82, January
[9] Leray J (1984) J Acta Math 63:193–248, 1934
[10] Guiraud JP, Zeytounian RKh (eds) (1986) Asymptotic modelling of fluid flows. Special Issue JTAM 1986
[11] Zeytounian RKh (1986) Les modèles asymptotiques de la mécanique des fluides, vol. I, LNP 245. Springer, Berlin
[12] Zeytounian Rkh (1987) Les modèles asymptotiques de la mécanique des fluides, vol. II, LNP 276. Springer, Berlin
[13] Zeytounian RKh (2002) Theory and applications of non-viscous fluid flows. Springer, Berlin
[14] Zeytounian RKh (2004) Theory and applications of viscous fluid flows. Springer, Berlin
[15] Kaplun S, Lagerstrom PA (1957) Asymptotic expansion of Navier-Stokes solution for small Reynolds numbers. J Math Mech 6:585–594
[16] Van Dyke M (1964) Perturbation methods in fluid mechanics. Academic Press, New York
[17] Van Dyke M (1975) Perturbation methods in fluid mechanics. Parabolic Press, Stanford
[18] Van Dyke M (1962) Second-order compressible boundary-layer theory with application to blunt bodies in hypersonic flow. In: Ridell Fr (ed) Hypersonic flow theory, vol. 7, Academic Press, New York, pp 37–76
[19] Guiraud JP, Zeytounian RKh (1971) Sur la structure des écoulements tourbillonnaires dans les turbomachines axiales. La Recherche Aérospatiale 2:65–87
[20] Guiraud JP, Zeytounian Rkh (1971) Sur la structure des écoulements tourbillonnaires dans les turbomachines axiales: Analyse de l'écoulement à l'entrée ou à la sortie d'une roue. La Recherche Aérospatiale 5:237–256
[21] Zeytounian RKh (1976) La météorologie du point de vue du mécanicien des fluides. Fluid Dyn Trans 8:289–352
[22] Zeytounian Rkh (1990) Asymptotic modelling of atmospheric flows. Springer, Berlin
[23] Zeytounian Rkh (1994) Modélisation Asymptotique en Mécanique des Fluides Newtoniens. SMAI, Maths et Appl, vol. 15. Springer, Berlin
[24] Zeytounian RKh (2006) *Topics in hyposonic flow theory*. LNP 672, Springer, Berlin
[25] Zeytounian Rkh (2009) Convection in fluids: a rational analysis and asymptotic modelling. FMIA 90, Springer
[26] Prandtl L (1904) Über Flussigkeitsbewegung bei sehr kleiner Reibung. In: Proceedings III—Int Math Congress, Heidelberg, pp 484–491
[27] Kibel IA (1940) Application to meteorology of the equations of hydromechanics. Izv Acad Sci SSSR, Geogr Geophys, No. 5
[28] Kaplun S (1957) Low Reynolds number flow past a circular cylinder. J Math Mech 6:595

# Chapter 1
# The Fluid Dynamics Adventure, from Newton to Stokes via Cauchy, Euler, Navier, Poisson, and Fourier

Our business below is the *dynamics of fluids* treated as *continua*—especially in *gases*, but also in *liquids*—and we shall need to know their *thermodynamic properties*. Here it seems interesting to note the following point. *Newton's* conception of fluid mechanics armed his successors for the next three centuries with basic tools which they used to erect the great structure of modern engineering science. Modern aerodynamics is completely Newtonian in its development. Newton postulated a special *frictionless incompressible* fluid, then a *viscous* fluid, and finally, a *compressible* fluid. From the concept of the frictionless, *ideal, and non-viscous* fluid comes the classical *hydrodynamics* of *Bernoulli*, *Euler*, d'*Alembert*, and *Lagrange*. From the concept of the *viscous fluid* we have first the *Navier incompressible and viscous equation*, and *Stokes'* and *Reynolds'* theories leading to *Prandtl's boundary layer*. From the concept of *compressibility*, we are right now evolving a comprehensive mechanics of *supersonic flow* to cope with the engineering needs of modern flight, since the Second World War.

In Sect. 1.4, the reader will find *various remarks* concerning the *Newtonian fluid, the Navier incompressible viscous equations, turbulence, Stokes' essays, the Prandtl boundary layer concept, and asymptotics.*

## 1.1 From Newton's Fundamental Equation to the Cauchy Stress Principle and the Macroscopic Equations of Continuum Mechanics

Fluid dynamics is a sub-discipline of fluid mechanics that deals mainly with fluid flow phenomena, in particular, in moving liquids (water) and gases (air). The fundamental axioms of *classical Newtonian fluid dynamics* in this book are the macroscopic *conservation laws of continuum mechanics* (see, for instance, Segel [1]). These are conservation of mass, linear momentum (from Newton's second law


R.Kh. Zeytounian, *Challenges in Fluid Dynamics*,
DOI 10.1007/978-3-319-31619-2_1

of motion), and energy (as in the first law of thermodynamics). The continuum assumption considers fluids to be continuous, rather than discrete (as is assumed in the kinetic theory of gases, a microscopic point of view). This continuum assumption is obviously a realistic view, if we take into account the fact that the macroscopic length scale L of various typical fluid flow phenomena in our environment is very large in comparison with the *mean free path* ($l$, a micoscopic length scale, the average distance between two successive collisions for one gas molecule picked at random, as is so well described in *Golse* [2], cited in "Preface"), i.e., $l/L \ll 1$, in other words the *Knudsen number* $Kn = l/L$ is a very small dimensionless parameter!

Here, I give a concise idea, but in a rational way, of the derivation of equations of continuum mechanics, according to Newton and Cauchy. In our 2001 survey paper [2], the reader can also find many references concerning the "*adventure*" *from Newton to Stokes!* It follows to begin with that we can use the *Cauchy stress principle* to write

$$\rho D\mathbf{u}/Dt = \rho \boldsymbol{f} + \nabla.\underline{\mathbf{T}}, \tag{1.1}$$

which is the equation of motion for the *velocity* vector $\mathbf{u}(t, \mathbf{x})$. In (1.1), the derivative along trajectories is $D/Dt = \partial/\partial t + \mathbf{u}.\nabla$, where the Cartesian components of the *gradient* operator $\nabla$ are $\partial/\partial x_j$, with $j = 1, 2, 3$, $\boldsymbol{f}$ is the *external body force* per unit mass and $\rho(t, x)$ the *density*, while $\underline{\mathbf{T}} = (T_{ij})$ is the second-order surface *stress tensor*, the components $T_{ij}$ being the coefficients of a matrix, and $T_{ij}$ the j-component of the force on the surface element with outer normal in the i-direction. Finally, t is the time variable, and $\mathbf{x} = (x_j)$ is the *position vector* in the fluid flow domain $\Omega$ in three-dimensional space.

Equation (1.1) is the simple and elegant equation of motion discovered by *Cauchy, 1828*, a rather more explicit form of the famous equation discovered by *Newton*, which can be written in the form

$$\rho D\mathbf{u}/Dt = \rho \boldsymbol{f} + \text{internal } \mathbf{force} \text{ per unit volume}, \tag{1.2}$$

as a direct consequence of his second law.

The above *Cauchy* equation (1.1) is valid for any fluid (i.e., gas or liquid, that is capable of *flowing freely*), regardless of the form which the (second-order) surface stress tensor, $\underline{\mathbf{T}}$ in (1.1), may take.

A *second consequence* of *Newton's principles* is then just a statement of the mass conservation law which, according to *Euler 1755*, can be written in the form:

$$D/Dt\left\{\int \rho\, dV\right\} = 0 \Rightarrow (1/\rho)D\rho/Dt + \nabla.\mathbf{u} = 0, \tag{1.3}$$

where dV is a volume element in the neighbourhood of the point $\boldsymbol{P}$ ( in $\Omega$, with coordinates $x_j$) and a mass $\rho dV$ is assigned to this volume element.

A (*homogeneous*) fluid is said to be incompressible if the density of any element remains constant as the time changes during the motion considered, i.e.,

$$D\rho/Dt = 0, \tag{1.4}$$

whence the equation of continuity for an incompressible fluid is simply

$$\nabla.\mathbf{u} = 0. \tag{1.5}$$

The vector $\nabla \times \mathbf{u} = \omega = \text{curl } \mathbf{u}$ is called the *vorticity*. If we suppose that $\omega = 0$ throughout any region, then a velocity potential $\varphi$ exists such that $\mathbf{u} = \nabla\varphi$, and if in addition the fluid is incompressible, we have

$$\nabla.\mathbf{u} = \nabla^2\varphi = 0, \tag{1.6}$$

where $\nabla^2$ is the 3D *Laplace operator* $\Delta_3 = \partial^2/\partial x_k \partial x_k$, with $k = 1, 2, 3$.

Concerning the conservation of energy, we first assume the existence of the usual *thermodynamic variables*, assuming also that the relations among them provided by classical reversible thermodynamics are satisfied. According to *Neumann 1894*, the equation of energy balance is written (per unit volume) as

$$\rho\left[De/Dt + p\,D(1/\rho)Dt\right] = \Phi + \nabla.\mathbf{q} + \rho\,R, \tag{1.7}$$

where p is the *thermodynamic pressure*, e the *internal energy*, and the term, $\left\langle\frac{De}{Dt} + p\frac{D}{Dt}\left(\frac{1}{5}\right) = 0\right\rangle$ the rate of *heat supply* needed for a fluid element, per unit mass. The second and third terms on the right of (1.7) give the heat supplied by conduction (via the *heat conduction vector* **q**) and *radiation* (R). Thus $\Phi$ represents the *heat gained* by the *dissipation of mechanical energy by internal friction*, the so-called *viscous dissipation function*. This agrees well with the result that $\Phi$ is a linear function of the $\tau_{ij}$, which are the components of the *shear stress tensor* (a *deviation* from the isotropic thermodynamic pressure, which *must be zero* when the fluid is *at rest*—see (1.28) below) due to the combined results of internal friction and distortion.

In *Goldstein 1960*, the reader will find a *tentative expression,* in invariant vector form, for the term $\Phi$—the *dissipation function*—in (1.7). We observe that the left side of the above energy equation (1.7) may be transformed into various forms by use of the equations of thermodynamics.

The most important involve the temperature, T, and the *entropy*, *S*, per unit mass. Hence we have the following very compact equation, showing the conditions for an isentropic (*non-viscous!*) fluid flow:

$$\rho T\,DS/Dt = \Phi + \nabla.\mathbf{q} + \rho\,R, \tag{1.8a}$$

and also two other forms of this energy equation for the temperature T:

$$\rho\,\mathrm{Cp}\,\mathrm{DT/Dt} + (\mathrm{T}/\rho)\,(\partial\rho/\partial T)\mathrm{p}\,\mathrm{Dp/\,Dt} = \Phi + \nabla.\mathbf{q} + \rho\,\mathrm{R}, \tag{1.8b}$$

$$\rho\,\mathrm{Cv}\,\mathrm{DT/Dt} + \mathrm{T}\nabla.\mathbf{u}(\partial\mathrm{p}/\partial\mathrm{T})\rho = \Phi + \nabla.\mathbf{q} + \rho\,\mathrm{R}. \tag{1.8c}$$

We observe that, for a (*thermally*) *perfect gas*, we may write the equation of state in the form

$$\mathrm{p} = \mathrm{R}\rho\mathrm{T}, \tag{1.9}$$

where R in (1.9) is an absolute constant divided by the molecular weight of the gas. For such a perfect gas, when the equation of state (1.9) is taken into account, the equations (1.8b) and (1.8c) reduce to:

$$\begin{aligned}\rho\,\mathrm{Cp}\,\mathrm{DT/Dt} - \mathrm{Dp/Dt} &\equiv \rho\,\mathrm{Cv}\,\mathrm{DT/Dt} + \mathrm{p}\nabla.\mathbf{u}\\ &= \Phi + \nabla.\mathbf{q} + \rho\,\mathrm{R}.\end{aligned} \tag{1.10}$$

Care must be taken for the motion of a gas, even at *low speeds*, when the fractional changes in density are not large, since the thermodynamic pressure p is usually *quite large*. In fact, as we shall see later, in such cases there are important problems in which it is Dp/Dt (and also $\Phi$) which has to be neglected, not $\mathrm{p}\nabla\cdot\mathbf{u}$ (as is the case for a truly *incompressible* fluid), and the operative specific heat is Cp, not Cv. The above equations (1.1), (1.3), and (1.7) for **u**, ρ, and e, viz.,

$$\begin{aligned}\rho\mathrm{D}\mathbf{u}/\mathrm{Dt} &= \rho f + \nabla.\underline{\mathbf{T}},\\ \mathrm{D}\rho/\mathrm{Dt} + \rho\nabla.\mathbf{u} &= 0,\\ \rho\mathrm{De/Dt} + \mathrm{pD/D}(1/\rho) &= \Phi + \nabla.\mathbf{q} + \rho\,\mathrm{R},\end{aligned} \tag{1.11}$$

are the general equations of *macroscopic classical continuum mechanics*.

Equation (1.8a) leads to an *elegant expression for the specific entropy S*—expressing the *rate of change of entropy following a particle—when it is assumed that* $\Phi = 0$, $\nabla.\,\mathbf{q} = 0$, and $\mathrm{R} = 0$. In fact, we obtain

$$\rho\mathrm{T}\ \mathrm{D}S/\mathrm{Dt} = 0 \Rightarrow \mathrm{D}S/\mathrm{Dt} = 0. \tag{1.12}$$

On the other hand, the thermodynamical interpretation of the internal energy e must naturally be somewhat different for *compressible* and *incompressible fluids*. For a *compresssible fluid* we assume that e is a *thermodynamic state variable*—as in Serrin [2] cited in "Introduction"—satisfying the thermodynamic relation

$$\mathrm{T}\ \mathrm{d}S = \mathrm{de} + \mathrm{p}\ \mathrm{d}(1/\rho). \tag{1.13a}$$

An important point about this relation, in the case of a *compressible fluid,* is linked with the fact that the *pressure*, as a *thermodynamical variable*, *must be* a function of ρ and *S*:

$$p = P(\rho, S). \tag{1.13b}$$

In fact, for an *incompressible fluid*, instead of (1.12 and 1.13a), it is assumed only that

$$T\, dS = de, \ \rho = \text{const}, \ e = E(S), \ \text{and } T = dE/dS, \tag{1.13c}$$

and the pressure *does not enter into the thermodynamical treatment of an incompressible fluid*. In Sect. 1.2.2 of Sect. 1.2, this fact will become clear for the reader! But for the consistency of the thermodynamics, the following restrictions must be satisfied:

$$\Phi - \mathbf{q}.[(\nabla T)/T] \geq 0 \Rightarrow \Phi \geq 0 \text{ and } \mathbf{q}.\nabla T \leq 0. \tag{1.13d}$$

Now, a crucial problem, for our main unknown useful functions, **u**, ρ, and T, is to obtain from the above system of equations (1.10) governing macroscopic continuum mechanics a closed *NSF system of equations*, as in our book, 2004 [14, pp. 14–25], cited in "Introduction". For this we need five complementary equations, in particular, for second-order stress tensor $\underline{\mathbf{T}}$, thermodynamic pressure p, temperature T, and heat flux vector **q** (see Sect. 1.3 below).

In real, *compressible, viscous and heat conducting*—Newtonian—classical fluids, a certain relation exists among the *three thermodynamic variables*, p, ρ, and T, i.e.,

$$f(p, \rho, T) = 0, \tag{1.14}$$

the so-called *equation of state for a trivariate fluid in baroclinic motion*.

## 1.2 Euler Non-viscous Equations and Navier Viscous Incompressible Equations

### *1.2.1 Euler Non-viscous Case*

*First*, concerning the (second-order) stress tensor $\underline{\mathbf{T}}$, which is a *symmetric* tensor for a fluid at rest, where there are *no shear stresses:* the only force acting on a material element of fluid is the *static pressure*, which acts *normally at very point on the surface of that element*. We can thus assume that

$$T_{ij} = -p\delta_{ij}, \tag{1.15}$$

where the $\delta_{ij}$ are the (Cartesian) components of the unit matrix $\underline{\mathbf{I}}$. This is the case for an *Eulerian non-viscous adiabatic (non-heat conducting!) perfect fluid flow*, when just Φ, **q** and R are all zero. The equations of a *compressible perfect, non-viscous,* *non-heat conducting Eulerian fluid flow,* first considered by *Leonhard Euler 1755*, are the following system of three equations:

$$\begin{aligned} \mathrm{D}\rho/\mathrm{Dt} + \rho\nabla.\mathbf{u} &= 0,\\ \rho\mathrm{D}\mathbf{u}/\mathrm{Dt} &= \rho f - \nabla\mathrm{p},\\ \mathrm{D}S/\mathrm{Dt} &= 0, \end{aligned} \tag{1.16}$$

for the unknown functions, ρ, **u**, and *S*—the entropy *S* being *conservative along the fluid trajectories*!

Since *tangential stresses* are neglected in the definition of a perfect (non-viscous) fluid, it is logical to suppose also that $\mathbf{q} = 0$—indeed, according to *kinetic theory,* the *viscosity* and *heat conduction* arise from *similar* mechanisms (*molecular impact*), and are of the *same order of magnitude!* Thus, if one is to be neglected, we should also neglect the other, and accordingly we have taken into account the above reduced equation (1.12) for the specific entropy *S*.

In the above Eulerian system of equations, (1.15), we also have to omit the radiation term (R), which plays an important role in the NSF equations for atmospheric motion (see Sect. 2.6).

The system of three equations (1.16) is a closed system for **u**, ρ, p, and *S* only when completed by a thermodynamical equation of state relating p, ρ and *S*. Hence, instead of the above equation of state (1.14), for an ideal (non-viscous) gas with constant specific heats $\mathrm{Cp} > 0$ and $\mathrm{Cv} > 0$, such that $\mathrm{Cp/Cv} = \gamma$ is a constant with value 1.40 for dry air, we can write

$$\mathrm{p} = \exp(S/\mathrm{Cv})\rho^{\gamma} \text{ with } \gamma = \text{ const } > 1,\ \mathrm{Cv} = \mathrm{R}/(\gamma - 1), \tag{1.17}$$

because $\mathrm{D}S/\mathrm{Dt} = 0$ in (1.16). The system of equations (1.16) above with (1.17), i.e.,

$$\begin{aligned} \mathrm{D}\rho/\mathrm{Dt} + \rho\nabla.\mathbf{u} &= 0,\\ \rho\mathrm{D}\mathbf{u}/\mathrm{Dt} &= \rho\mathbf{f} - \nabla\mathrm{p},\\ \mathrm{D}S/\mathrm{Dt} &= 0,\\ \mathrm{p} &= \exp(S/\mathrm{Cv})\rho^{\gamma},\\ \gamma = \mathrm{Cp/Cv} &= \text{ const } > 1, \end{aligned} \tag{1.18}$$

is the *complete Euler compressible, non-viscous and non-heat conducting system of equations governing* **u**, ρ, *S*, and p *for an ideal gas.*

For an Eulerian *incompressible,* but *nonhomogeneous fluid flow,* a so-called *isochoric fluid flow*, when we assume that the exterior force ***f*** is the *gravitational force* **g** per unit mass, instead of the above system (1.17) of Euler compressible equations, the following more simple system for **u**, ρ, and p is often useful:

$$\begin{aligned} \nabla \cdot \mathbf{u} &= 0, \\ \rho\, \mathbf{Du}/\mathrm{Dt} &= \rho \mathbf{g} - \nabla \mathrm{p}, \\ \mathrm{D}\rho/\mathrm{Dt} &= 0. \end{aligned} \tag{1.19}$$

In this case, the density ρ is conservative along the fluid trajectories (rather than the entropy $S$). In the book by Yih [3], such a system (1.18) of isochoric equations is considered for the investigation of *"stratified flows"*, which have various applications in *meso-meteorology,* in particular, to *lee wave phenomena downstream of a mountain*, considered in Sect. 8.6 of the Chap. 8.

It is interesting to note that, in *Euler 1755*, only the *barotropic case* was actually considered, when the specific entropy $S = \text{const}$, so that in the above system of equations (1.18) for p, the following *particular specifying relation* is assumed:

$$\mathrm{p} = \mathrm{P}(\rho). \tag{1.20a}$$

In fact, a closed system is derived for **u**, ρ, and p alone, when the following two equations are considered with (1.20a):

$$\mathrm{D}\rho/\mathrm{Dt} + \rho \nabla . \mathbf{u} = 0, \text{ and } \rho\, \mathbf{Du}/\mathrm{Dt} = \rho f - \nabla \mathrm{p}. \tag{1.20b, c}$$

*In 1755, 260 years ago, using the above formulation* (1.20a–c), *Euler believed that he had reduced fluid dynamics, in principle, to a branch of mathematical physics, and this was obviously a very striking achievement in 1755 when fluid dynamics was still in its infancy!*

But progress in integrating the above system (1.18)—the Euler system of equations for a compressible ideal gas—was slow at first. This is well expounded in *Birkhoff's 1983 survey paper* [4], cited in "Introduction", which briefly reviews the *gradual disintegration of Euler's concept of analytical fluid dynamics as a mathematical science.*

For instance, according to Birkhoff's survey:

> In *1781, Lagrange* gave the first proof of the permanence of irrotational flow—but a rigorous demonstration (in *1815*) is due to *Cauchy* and published in *1827*. From *1840* to *1915*, the study of *potential* (*irrotational* and *incompressible*) *flows* was the most active branch of *analytical* (*à la Lagrange*) *fluid mechanics, and the theory of potential flow,* with a *free surface at constant pressure* (*waves in water*—see Sect. 8.4)—also developed brilliantly, and in the review paper by Zeytounian [4]—dedicated by the Editorial Board of Uspekhi Fizicheskikh Nauk (Russian Academy of Sciences) to the *30th anniversary* of the publication of the paper by *Zabuski* and *Kruskal, 1965*, in which the term *soliton* was *mentioned for the first time* in the scientific literature—the reader can find some of the main points of this development.

In particular, the book by Newell [5] gives an excellent, attractive, and exciting view of the theory of solitons, which brings together many branches of mathematics, some of which touch upon profound ideas! On the other hand, there is the famous d'*Alembert paradox*: *A solid moving through an ideal fluid, as governed by the Euler equations, would encounter no resistance at all?* This d'*Alembert paradox* is also related to the so-called *Kutta, 1902 and Joukowski, 1912* condition, which is closely connected with the *well-posedness of the fluid dynamics equations* and the *behavior of fluid flows with vanishing viscosity.*

Indeed, according to the famous Stewartson [6] paper, the mathematical conjectures associate with d'*Alembert's paradox* can be stated *most clearly* in terms of (see below) the *Navier equation* (1.25) for an *incompressible viscous fluid* (see, for instance, Sects. 4.1.2 and 4.2.2 of our 2014 "*Scientific Autobiography*").

In the long paper (50 pages) Guiraud [7], which is rather a *collection of thoughts on the role of asymptotics in a variety of fields*, the reader can find (pp. 269–271) some comments concerning d'Alembert's paradox, and also various comments concerning the *asymptotics of the triple deck* (the early roots of the theory, the *upstream influence* mystery, *Lighthill's* theory, the main *features of asymptotics,* and a *few examples of significant achievements*). In Chap. 12 of our *2002 monograph, FMIA 64* [7, pp. 471–525 and 542–545], cited in "Introduction", the reader can find a detailed account of *singular coupling and the triple-deck model.*

Section 1.4, the reader can find some complementary information concerning *Euler non-viscous fluid flows,* and *our two books*: *2002* [13], cited in "Introduction", and *1974* [8], are both devoted to various applications of *Eulerian fluid flows.*

Here, we observe that, (*non-viscous*) fluid dynamics was first envisaged as a systematic branch of *mathematical physics* in *Johann Bernoulli*'s, *1737* "*Hydraulics*" and in *Daniel Bernoulli*'s *1738* "*Hydrodynamica*". *Fluid dynamics was subsequently expounded in d'Alembert*'s *1744* treatise "*Traité de l' équilibre et du mouvement des fluides*". *Johann Bernoulli*, was then regarded as Europe's foremost mathematician, and would eventually be the most important influence on the young *Leonhard Euler*'s formal education, which began in Basel, where he was sent to live with his maternal grandmother. In *1720*, aged 13, he enrolled at the *University of Basel*, and in *1723*, he received a *Master of Philosophy.*

At that time, he was receiving Saturday afternoon lessons from *Johann Bernoulli*, who quickly discovered his new pupil's *incredible talent for mathematics!* Euler studied various fluid dynamics problems, and in the mid-1700s, *formulated the fundamental ideas expounded in the three books mentioned above, mathematically, as partial differential equations in the ground-breaking paper Euler, 1755*. It is firmly established that *Euler was indeed the founder of rational fluid mechanics.*

The above *Euler system of equations* (1.18) appear to be *very complex*, but they are actually considered—in the framework of the RAM approach—only as a *simplifications/reduction of the more general NSF equations* (which are really *considerably more complicated*!) to a *non-viscous and adiabatic* fluid flow.

The Euler equations *neglect the effects of the viscosity of the fluid*—this viscosity effect being included in the *Navier equation* (see (1.26a) in Sect. 1.2.2), and this viscosity effect is also considered in the NSF equations via the Stokes "*fluidity concept*" (se Sect. 1.3). *In the framework of the present book,* solution of the Euler equations is therefore *only an approximation to a real fluid flow situation! Still the fact remains that, for some problems, like the lift of a thin airfoil at low angle of attack, a solution of the Euler equations provides a good model of reality. But for other problems, like the growth of the Prandtl boundary layer on a flat plate, the*

*Euler non-viscous equations obviously do not properly a model problem, when we consider the initial and boundary conditions.*

In the MAE framework, one must consider the *Euler-Prandtl regular coupling discussed* in Sect. 5.4.1 of the companion 2012 book, and with more detail in Sect. 6.1, where the reader will find a deeper investigation in the context of *large Reynolds* number.

Below, as an *analytical application* relating to the Euler equations, we briefly consider the case of a *steady 3D Euler inviscid compressible fluid flow*, and give a new and well adapted form for the above Euler system of equations (1.18). This new form of the *Euler steady 3D* equations was discovered in Moscow in 1965 by me (50 years ago!) and published in 1966 [9], thanks to *A. M. Obukhov*, who was the director of the *Institute of Atmospheric Physics* (IFA) in Moscow in *1966*, well known for the famous "*Kolmogorov-Obukhov law*" *for turbulence*. This new form is well suited for the investigation of *3D turbomachinery flows* (considered in Sect. 8.2), and also the *lee waves downstream of a mountain* (considered in Sect. 8.6). Let us suppose that the external force is conservative, so that $\boldsymbol{f} = -\nabla U$, and assume that the flow is *steady* ($\partial/\partial t = 0$), whence

$$\mathrm{D}\mathbf{u}/\mathrm{Dt} \equiv \omega \times \mathbf{u} + (½)\nabla|\mathbf{u}|^2, \text{ and } \nabla.(\rho\mathbf{u}) = 0. \tag{1.21a, b}$$

Introducing the *specific enthalpy* $h = e + p/\rho$, after a little computation, we get the so called *Vazsonyi form of the Euler steady-state equation for* **u**:

$$\mathbf{u} \times (\nabla \times \mathbf{u}) = \nabla H - T\nabla S \text{ with } H = h + (½)|\mathbf{u}|^2 + U, \tag{1.21c, d}$$

such that:

$$\mathbf{u}\,.\,\nabla H = 0 \text{ and } \mathbf{u}\,.\,\nabla S = 0. \tag{1.21e, f}$$

Consequently, *both surfaces* $H = \text{const.}$ (*Lamb* surface) and $S = \text{const.}$ (*isentropic* surface) are *stream surfaces* in the considered steady-state 3D Eulerian compressible, baroclinic fluid flow, and we can introduce two sets of stream functions in 3D space: $\psi(t, \mathbf{x})$ and $\chi(t, \mathbf{x})$, such that $\mathbf{u}\,.\,\nabla\psi = 0$ and $\mathbf{u}\,.\,\nabla\chi = 0$. The intesection of these two stream surfaces, $\psi(t, \mathbf{x}) = \text{const.}$ and $\chi(t, \mathbf{x}) = \text{const.}$, in 3D space give the *streamlines of three-dimensional steady-state Eulerian inviscid and compressible fluid flow*.

Hence, as a direct consequence of the above steady-state continuity equation (1.21b), we can write the following 3D representation for the velocity vector **u**:

$$\mathbf{u} = (1/\rho)(\nabla\psi \times \nabla\chi). \tag{1.22a}$$

On the other hand, according to (1.21e,f), we can write the following relations for H and S:

$$\mathrm{H} = \mathrm{H}(\psi, \chi) \text{ and } S = S(\psi, \chi), \tag{1.22b, c}$$

where the functions H(ψ, χ) and S(ψ, χ) are *both conservative along the streamlines of the 3D Eulerian compressible baroclinic steady fluid flow*.

The next decisive step is to substitute **u** of (1.22a) into the Vazsonyi equation (1.21c).

We thus obtain the following *rather awkward* vector equation:

$$\begin{aligned}&(\nabla\psi \times \nabla\chi) \times (\nabla \times \mathbf{u})\\&= \rho\{(\partial \mathrm{H}/\partial\psi)\nabla\psi + (\partial \mathrm{H}/\partial\chi)\nabla\chi - \mathrm{T}\,[(\partial S/\partial\psi)\nabla\psi - (\partial S/\partial\chi)\nabla\chi]\},\end{aligned} \tag{1.23a}$$

or

$$\begin{aligned}&[(\nabla \times \mathbf{u}).\nabla\psi]\nabla\chi - [(\nabla \times \mathbf{u}).\nabla\chi]\nabla\psi\\&= \rho\{\partial \mathrm{H}/\partial\psi - \mathrm{T}\partial \mathrm{S}/\partial\psi\}\nabla\psi + \rho\{\partial \mathrm{H}/\partial\chi - \mathrm{T}\partial \mathrm{S}/\partial\chi\}\nabla\chi.\end{aligned} \tag{1.23b}$$

In fact, this last vector equation (1.23b) gives *two scalar equations*:

$$(\nabla \times \mathbf{u}).\nabla\psi = \rho\{\partial \mathrm{H}/\partial\chi - \mathrm{T}\partial \mathrm{S}/\partial\chi\}, \tag{1.24a}$$

$$(\nabla \times \mathbf{u}).\nabla\chi = -\rho\{\partial \mathrm{H}/\partial\psi - \mathrm{T}\partial \mathrm{S}/\partial\psi\}. \tag{1.24b}$$

The above two "*first integrals*", (1.24b) and (1.24a), with (1.22e,f) for **u** and the relations (1.22b,c) for H(ψ, χ) and S(ψ, χ), give a *new analytical formulation* of *Eulerian non-viscous steady 3D compressible fluid flow!*

Unfortunately, up until now, *I am not sure whether I was the first to discover the above two first integrals* (1.24b) and (1.24a) *in Moscow (1965)*. However that may be, to the best of my recollection, *Kibel* was somewhat *surprised*, but had a very positive reaction, especially because some time ago he showed me *a reduced/simplified form* of the above equations (1.24b) and (1.24a) for a simple steady incompressible 2D case!

As I wrote in my *Scientific Autobiography, 2014* (p. 38), the discovery of the two first integrals (1.24b) and (1.24a) had various important consequences for me, and in particular opened new options for me regarding my theoretical research on fluid dynamics during my *second scientific career,* in the years 1967–1996 at ONERA and at the University of Lille I in France.

To conclude, I wish to add that the interested and motivated reader can find in [13], cited in "Introduction" and [8] various discussions relating to non-viscous fluid flows. In particular, concerning my 1974 book [8], *S. Childress* in *J. Fluid Mechanics* **78**, part 1, 207–208, of 1976, wrote:

> The first five chapters comprise a through summary of the classical general theory and its recent extensions, and the last five chapters treat applications to some specific problems. Of

particular interest in the former is the powerful use of the Lagrangian description, not only in Chapter II on material invariants, but also in the kinematic and dynamic analysis of adiabatic baroclinic flow given in chapter III. Chapter IV treats in detail the question of compatibility between velocity and force fields, that is, conditions under which the given fields determine a forced perfect fluid flow. This chapter is overly technical and really not needed in the later applications. In Chapter V an expanded discussion of variational principles is given. In fact, the heart of the remainder of the book is Chapter IX on the asymptotic theory of axial flow through a turbine, which will likely be of considerable interest to specialists—the analysis is a fascinating unravelling of a complicated engineering problem using asymptotic methods with the basic large parameter being the number of turbine blades per rotor. The remaining chapters are concerned with linearized flow, asymptotic models, stationary flows, and meteo applications. For example, in anticipation of Chapter IX, one linearization describes three-dimensional perturbations of a helicoidal flow. The asymptotic methods given in Chapter VII are restricted to low Mach number limit process meteo problems. Especially noteworthy among the meteo problems is the discussion of three-dimensional lee waves, with various applications not readily found elsewhere!

With regard to my 2002 [13], cited in "Introduction" book on the *theory and applications of non-viscous fluid flows,* the reader will find *three chapters* entitled *Fluid Dynamic Limits of the Boltzmann Equation, Newtonian Fluids and NSF Equations,* and *A short Presentation of Asymptotic Methods and Modelling.* In Chap. 4, we begin with a formulation and study of *various forms of Euler equations and also of some problems arising in hydro-aerodynamics (the d'Alembert paradox and the Kutta-Joukowski-Villat condition, potential flow and water waves, isochoric fluid flow, and transonic and hypersonic gas dynamics model equations).* Chapter 5 is devoted to *atmospheric baroclinic motions,* and in particular we give an asymptotical derivation of the *famous inviscid Boussinesq equations,* which are very useful for predicting lee-wave phenomena downstream of a mountain. Chapter 6 deals with *low Mach number flows* and provides a way to derive *linear acoustic (initial/inner) equations associated with incompressible Euler (outer) equations; more precisely, the low Mach number asymptotics of weakly compressible Euler fluid flow is doubly singular near the initial time* $t=0$ (where the initial data are imposed) and also *at infinity, when the space variable* $\|\mathbf{x}\| \to \infty$, because of the singular behavior of the classical *Janzen-Rayleigh* proximate (near field) expansion! Chapter 7 sketches *an asymptotic theory of turbomachinery flow (within a row and at the inlet/outlet of this row), according to work by Guiraud and Zeytounian.* Chapter 8 is devoted to *vortex sheets* and *shock layer phenomena,* and in particular, we describe the *Guiraud-Zeytounian theory for "rolled-up" vortex sheets and the structure of the Taylor shock layer.* Finally, in Chap. 9, the reader can find some recent *rigorous results concerning the mathematical theory of non-viscous fluids.* In particular, for *incompressible non-viscous fluids,* we mention the very pertinent book by Marchioro and Pulvirenti [10], based on sound documented evidence.

### 1.2.2 Navier Viscous Incompressible Case

Concerning the application of the Cauchy stress principle (1.1), for a more realistic fluid flow, we stress that it was not until 1821, 66 years after Euler, but 20 earlier than Stokes, that *Navier was the first to take into account viscosity.*

Using a molecular model, he obtained for an incompressible homogeneous fluid flow, as a consequence of the Cauchy stress principle (1.1), *instead of the above Euler equation for a compressible non-viscous motion* for **u** in (1.15), viz.,

$$\rho\, \mathbf{Du}/\mathrm{Dt} = \rho \boldsymbol{f} - \nabla \mathrm{p}, \tag{1.25}$$

the following new *viscous* but *incompressible equation for* **u**:

$$\mathbf{Du}/\mathrm{Dt} + (1/\rho^{\circ})\nabla \mathrm{p} = \boldsymbol{f} + \nu^{\circ}\nabla^2\mathbf{u}, \text{ with } \nabla\,.\,\mathbf{u} = 0, \tag{1.26a}$$

where $\rho^{\circ}$ is a constant density, $\nu^{\circ} =$ const. is the kinematic viscosity coefficient, and p is the pressure.

Navier was the first to develop a theory of suspension bridges, and he designed a suspension bridge over the River Seine in Paris which attracted public attention. But Navier is not remembered today as a famous builder of bridges, as he was in his own day, by rather for the above *Navier equation* (1.26a).

By analysing fluid movement, and assuming a force of repulsion and attraction between neighbouring molecules in addition to the force studied by Euler, *Navier found the equation of motion of a viscous fluid.* Thereafter, using research by Cauchy, Poisson, and Saint-Venant, Stokes derived the famous system of equations,—first called the *NSF equations* by me on p. 3 of my 2002 book [13], cited in "Introduction", including viscosity and heat conduction and governing compressible fluid flows (see also p. 1 and pp. 52–53 of my book edited in *2012*).

Note that, in the above Navier equation (1.26a), it is crucial to bear in mind that the term $(1/\rho^{\circ})\nabla \mathrm{p}$ *is not an unknown quantity* in the initial-boundary value Navier problem for **u**! This *force term acts on the particles* of an incompressible and homogeneous viscous fluid, which *move as freely as possible*, but *in a way compatible* with the *incompressibility constraint* $\nabla\,.\,\mathbf{u} = 0$.

Hence, thanks to the equation $\nabla\,.\,\mathbf{u} = 0$ and using the Navier incompressible viscous equation (1.26a), we derive (assuming $\boldsymbol{f} \equiv 0$) a *Poisson equation* with *two source terms*, proportional to $\rho^{\circ}$ and $\mu^{\circ} = \nu^{\circ}\rho^{\circ}$:

$$\Delta_3 \mathrm{p} = -\rho^{\circ}\nabla\,.\,[(\mathbf{u}\,.\,\nabla)\mathbf{u}] + \mu^{\circ}\nabla\,.\,(\Delta_3\mathbf{u}), \tag{1.26b}$$

where the pressure p is determined by the velocity vector field **u** using the incompressibility constraint $\nabla\,.\,\mathbf{u} = 0$.

In such a case, p is a *nonlocal functional* of the *instantaneous* (the speed of sound is effectively infinite!) *flow configuration via* **u**. Let us note that, assuming $\boldsymbol{f}$

conservative, we may directly obtain an equation for the vorticity $\omega = \nabla \times \mathbf{u}$, namely:

$$\partial\omega/\partial t + \nabla \times (\omega \times \mathbf{u}) = \nu^\circ \Delta_3 \omega, \text{ with } \nabla . \mathbf{u} = 0, \tag{1.26c}$$

where $\nabla^2 = \Delta_3$ is the 3D Laplacian operator.

But any potential flow $\mathbf{u} = \nabla\varphi$, for which we thus have $\omega = 0$, trivially satisfies the above equation (1.26c). However, when $\nu^\circ > 0$, it is essential to replace the *slip boundary condition*, $\mathbf{u} . \mathbf{n} = 0$, on the *fixed body wall* $\Gamma$, for a *non-viscous Euler fluid flow*, by the *more stringent no-slip* condition, $\mathbf{u} = 0$, on the fixed *body wall* $\Gamma$, for a *viscous Navier fluid* ($\mathbf{n}$ is the outer unit normal to $\Gamma$).

In pp. 122–123 of our *2012*companion book, the reader can read more about the mathematical nature of the Navier equation (1.26), and the formulation of a well posed initial-boundary problem *à la Ladyzhenskaya*. Indeed, *Olga Alexandrovna Ladyzhenskaya* (1922–2004) played a particularly *important role* here by publishing, during the years 1957–2002, numerous papers and books on the *"purely" mathematical aspect of "viscous, incompressible, unsteady Newtonian fluid flow"*. Significantly, there are many important results in her *2003* paper (in *Russian Mathematical Surveys*, vol. 58 (2), pp. 251–286), entitled "*Sixth problem of the millennium: Navier-Stokes equations, existence and smoothness*", and on p. 122 of the book 2012, the reader will find her *major theorem*. She even presents a list "*of what to prove in order to solve this above mentioned sixth problem—of the seven problems of the millennium"*, proposed on the internet at <http://www.Claymath./>.

For a *vanishing* $\nu^\circ$, the Navier equation (1.26a) reduces to the Euler equation for an incompressible ($\nabla . \mathbf{u} = 0$) homogeneous ($\rho = \rho^\circ = \text{const}$) and non-viscous fluid flow:

$$\mathrm{D}\mathbf{u}/\mathrm{Dt} + (1/\rho^\circ)\nabla p = f. \tag{1.26d}$$

But because $\nu^\circ > 0$ multiplies the *derivative of the highest order in the Navier equation* (1.26a), *it cannot be inferred that the solutions of the Navier equation* (1.26a), *for very small values of* $\nu^\circ$, *tend uniformly to solutions of the above incompressible Euler equation* (1.26d). The *viscosity* $\nu^\circ \downarrow 0$ is a *very singular limiting process, and leads to a typical application of the RAM approach, as explained in* Sect. 6.1.

With the no-slip condition $\mathbf{u} = 0$, the *fluid adheres to a fixed body wall* $\Gamma$, so that the (*total*) velocity there is zero, or equal to the velocity of the moving body when the body moves. *The Navier equation is a valuable fluid flow model equation up to the wall* $\Gamma$. If the velocity is small and *the path of the fluid along the wall not too long*, the *velocity will again have its usual value near the wall, outside of the very thin transition boundary-layer (BL).*

In this *transition BL*, in spite of the small viscosity coefficient $\nu^\circ$ (high Reynolds number $\mathrm{Re} = \mathrm{U}^\circ\mathrm{L}^\circ/\nu^\circ \gg 1$), the sharp changes in velocity produce noticeable effects. Prandtl [26], cited in "Introduction", not only mentions the existence and nature of this BL and its connection with frictional drag, but also derives heuristically the BL-Prandtl equations valid in the thin viscous layer very close to the wall of the solid body.

In the framework of Chap. 6, in Sect. 6.1, the *complicated nature* of the *unsteady concept of boundary layer* is revealed, thanks to the *high Reynolds number limiting case* Re$\gg$1, by the RAM approach applied to the *full unsteady NSF system of equations.*

Concerning the boundary-layer concept, it seems to me that one should not overlook the important investigations into the nature of the boundary layer by Lanchester (1907) in England, including his explanation of separation (independently of Prandtl who, in 1904, was also interested in separation phenomena).

For a two-dimensional fluid flow $\mathbf{x} = (x, y)$ and $\mathbf{u} = (u, v)$, if $\psi(x, y)$ is the stream function, then the Navier equation for $\psi$ takes the form

$$\partial/\partial t(\nabla^2\psi) - [\partial(\psi, \nabla^2\psi)/\partial(x, y)] = \nu^\circ \nabla^4\psi, \tag{1.27}$$

where $u = \partial\psi/\partial y$ and $v = -\partial\psi/\partial x$. There are a number of simple, but physically important solutions for which the *nonlinear term*, viz.,

$$[\partial(\psi, \nabla^2\psi)/\partial(x, y)] \equiv [(\partial\psi/\partial y)\partial(\nabla^2\psi)/\partial x - (\partial\psi/\partial x)\partial(\nabla^2\psi)/\partial\psi(\partial\psi/\partial y)],$$

in (1.27) is *identically zero*. Amongst the *steady motions*, the most important case is *Poiseuille flow—steady laminar flow through a circular tube under the influence of a pressure gradient*, sufficiently far from the ends of the tube for the velocity distribution across a section to be the same in all sections—the *velocity distribution is then parabolic*. Problems of unsteady motion, with quadratic term zero, are essentially problems in the diffusion of vorticity, for which the same mathematical methods may be used as for other linear diffusion problems, as in the solution of the heat-conduction equation. We mention also the flow due to a rotating disk; see for instance pp. 75–78 in Chap. 3 of [14], cited in "Introduction", devoted to some simple examples of viscous fluid flows. In Sect. 8.8 of the present book, we investigate only the *flow of a thin liquid film over a rotating disk.*

*Low Reynolds* number flow (*very viscous* fluid flow—Re $\ll$1) *past a circular cylinder* was first considered in the framework of asymptotics by *Kaplun* [28], cited in "Introduction". The reader can find further information in *Nickel (1973)*, entitled *Prandtl's boundary-layer theory from the viewpoint of a mathematician*. In particular, the so-called *zero-viscosity limit* of the Navier equation was recently considered by Weinan, 2000.

To conclude this section, it is also important to note a highly *undesirable and singular fact* concerning the *passage from compressible to incompressible flow*, which *filters out fast acoustic* waves. This "*zero-compressibility limit*" is also (like the "*zero-viscosity limit*") a *singular limiting process, and the reader should refer to the short but significant paper by Constantin, 1995*. In Sects. 6.1 *and* 6.2, these two singular limiting processes will be considered via the RAM approach.

In particular, the Navier equation (1.26a) is derived from the full unsteady NSF system of dimensionless equations, in the framework of a *low Mach number* (M, the ratio of a reference velocity to a constant speed of sound), $M \downarrow 0$, as a model (outer) equation. In this case, with the *Navier equation*, one must also take into account the *associated acoustic (inner) equation* which is significant *near time* $t = 0$, and consider an *unsteady adjustment-transient problem*, in the framework of a *full unsteady initial-boundary value NSF problem*. In addition, one must derive the "*low Mach number limit*" of the compressible equation for the *temperature T* (which is (1.10) for a *perfect gas*). One can then write a *thermal condition* for T on the body wall $\Gamma$. This temperature condition leads to a *similarity relation between the low Mach number and the dimensionless thermal parameter* for the dimensionless temperature condition.

The resulting system of equations—*Navier with acoustics and the reduced equation for the temperature T*—leads to the so-called *Navier-Fourier incompressible reduced model*.

Finally, one should *not overlook* the important investigations by *Poisson, 1831* who not only assumed (like Navier) a molecular model for the viscous incompressible case, but who also gave (in *1843)—before Stokes* (in *1845*)—the general expression for the stress tensor (see Sect. 1.3).

*During the years 1986/1987 to 2004, I have devoted much time to the* Navier equations ((1.26a) with $\nabla . \mathbf{u} = 0$), in my books published in *1991* (*Mécanique des fluides fondamentales,* Lecture Notes in Physics, *m4*), *1994* (*Modelisation asymptotique en mécanique des fluides newtoniens*, Mathématiques & Applications *15*, *SMAI*), *2002* (*FMIA 64*), and *2004* (*Theory and Applications of Viscous Fluid Flows*). *The reader can thus obtain a* broad overview of the different facets of applications of these Navier incompressible and viscous equations!

However, the reader must once again take into account the fact that the Navier equations *neglect effects due to the compressibility of the fluid—this compressibility effect with the viscous and heat conducting effects being* included in the NSF equations (see Sect. 1.3), *thanks to Stokes. As in the case of the Euler inviscid equations, solution of the Navier equations is therefore only a particular approximation to a real fluid flow situation* leading to the NSF equations!

## 1.3 Stokes' Concept of Fluidity: Cauchy-Poisson and Fourier Constitutive Laws

It is the *shear stress tensor which is responsible for change in shape, a fact which leads us to the following decomposition:*

$$T_{ij} = -p\delta_{ij} + \tau_{ij}, \tag{1.28}$$

where $\tau_{ij}$ are the components of the *shear stress tensor*. Such a decomposition must be made *in order to deal with the general case of a compressible, viscous and heat conducting, non-adiabatic fluid flow*.

We observe that *p is the thermodynamic pressure*, with (−) since the sign convention is that $(+)T_{ij}$ *is tension. In* (1.28), $\delta_{ij}$ selects only the normal components of the stress tensor, since *p* acts normal to a surface. The expression (1.28) for $T_{ij}$ is a *constitutive equation*—a relation for a *Newtonian fluid*, an idealized fluid, that approximates the behavior of *water*, *air*, *gas*, *liquids*, and indeed many other fluids!

### *1.3.1 Stokes' Four Postulates*

When he was only 26 years old, *Stokes stated the concept of fluidity*, and the reader can find in the 2004 book [14, pp. 14–25], cited in "Introduction", a detailed derivation of the *NSF* equations inspired by [2, pp. 230–239], cited in "Introduction". Note, however, that these *NSF equations* ignore the name of the only person they should not have forgotten, namely *Saint-Venant*; see *A. J. C. Barré de Saint-Venant, <Note à joindre à un mémoire sur la dynamique des fluides> in* C. R. Acad. Sci. 17 (*1843*) 1240–1244. This note is developed in < *Mécanique appliquée* de Navier—annotée par *Saint-Venant*, 1858>, where, after *Navier*'s death (in 1836), his former pupil demonstrated what Navier had found. *Saint-Venant*'s work was the beginning of the *mechanical approach to continua*, which was to last more than a century and a half.

In *1845*, *G. G. Stokes* demonstrated Navier's equations once again, using a somewhat different approach (see *J.D. Anderson*: *History of Aerodynamics*, Cambridge Univ. Press, *1997*). In fact this *Stokes concept of the fluidity* can be stated in the form of *four postulates:*

1. $\underline{\mathbf{T}}$ is a continuous function of $\underline{\mathbf{D}}(\mathbf{u})$ and is independent of all other kinematic quantities.
2. $\underline{\mathbf{T}}$ does not depend explicitly on the space position vector **x**—spatial *homogeneity*.
3. There is no preferred direction in space—isotropy.
4. When $\underline{\mathbf{D}}(\mathbf{u}) = 0$, then $\underline{\mathbf{T}}$ reduces to—p $\underline{\mathbf{I}}$.

A medium whose constitutive equation satisfies the above four postulates is called a *Stokesian fluid*. According to postulates *1* and *2*, *we can write*

$$\underline{\mathrm{T}} = f\left(\underline{\mathbf{D}}(\mathbf{u})\right), \tag{1.29a}$$

where $\underline{\mathbf{D}}(\mathbf{u}) = (d_{ij})$ is the *deformation tensor*.

### *1.3.2 Cauchy-Poisson Law for the Cauchy Stress Tensor*

If we add, to Stokes' above four postulates, the condition that the components of $\underline{\mathbf{T}}$ be *linear* in the components of $\underline{\mathbf{D}}(\mathbf{u})$, then $\underline{\mathbf{T}}$ must reduce to the following general form:

$$\underline{\mathrm{T}} = [-\mathrm{p} + \lambda(\nabla . \mathbf{u})]\underline{\mathbf{I}} + 2\mu\underline{\mathbf{D}}(\mathbf{u}), \qquad (1.29\mathrm{b})$$

as a *Cauchy-Poisson* constitutive law, where p is the thermodynamic pressure, and $\lambda$ and $\mu$ are two scalar functions of the thermodynamic state. With the usual Cartesian cordinates, we have the following formula for the components $\mathrm{d}_{ij}$ of $\underline{\mathbf{D}}(\mathbf{u})$:

$$\mathrm{d}_{\mathrm{ij}} = (1/2)\left[\partial \mathrm{u}_\mathrm{i}/\partial \mathrm{x}_\mathrm{j} + \partial \mathrm{u}_\mathrm{j}/\partial \mathrm{x}_\mathrm{i}\right]; \quad \mathbf{u} = (\mathrm{u}_\mathrm{k}), \mathrm{k} = 1, 2, 3. \qquad (1.29\mathrm{c})$$

For an *incompressible fluid,* with $\mu$ as a single scalar function of temperature and $\nabla . \mathbf{u} = 0$, we have

$$\underline{T} = -\mathrm{p}\underline{\mathbf{I}} + 2\mu\underline{\mathbf{D}}(\mathbf{u}), \qquad (1.29\mathrm{d})$$

*and the pressure* p *in this particular incompressible case satisfies the Poisson equation* (1.26b).

The *constitutive equations*, viz., (1.29b) for the compressible case and (1.29d) for the incompressible case, constitute the *Cauchy-Poisson law*. These two classical/Newtonian *constitutive equations* first appeared in the work *of Poisson* (1831) entitles *Mémoire sur les équations générales de l'équilibre et du mouvement des corps solides élastiques et des fluides*. The formula for the *viscous dissipation function* $\Phi$ corresponding to the above *compressible Cauchy-Poisson law* (1.29b) is

$$\Phi = \underline{\mathbf{T}} : \underline{\mathbf{D}} = \lambda(\nabla.\mathbf{u})^2 + 2\mu\underline{\mathbf{D}} : \underline{\mathbf{D}}, \qquad (1.29\mathrm{e})$$

but the condition $\Phi \geq 0$, places some restrictions on the possible values of the viscous coefficients $\lambda$ and $\mu$. Indeed, for an *incompressible fluid* we must have $\mu \geq 0$, but for a *compressible fluid,* we must have $\mu \geq 0$ and $3\lambda + 2\mu \geq 0$, but here I do not discuss the various controversies concerning the famous Stokes case:

$$\lambda = -(2/3)\mu, \qquad (1.30)$$

which is called the *Stokes relation*. Nevertheless for a *monoatomic perfect gas*, when the *bulk viscosity coefficient* $\mu_v = \lambda + (2/3)\mu$ *vanishes*, this *Stokes relation is indeed satisfied*!

In fact, when written in component form, the relatively simple formula from (1.29b) and (1.29c) for the second-order *shear stress tensor* is, according to decomposition (1.28):

$$\tau_{ij} = \lambda(\partial u_k/\partial x_k)\delta_{ij} + \mu(\partial u_i/\partial x_j + \partial u_j/\partial x_i), \tag{1.31}$$

which, according to the concept of fluidity, is linearly proportional to the symmetric part of $d_{kl}$ (components of the second-order *deformation* tensor). This is mainly a consequence of the *isotropy constraint*: *there are no preferred directions in the fluid*, so the *fluid properties are point properties!*

### 1.3.3 Fourier Law for the Heat Flux Vector

Fourier's most famous work is *Théorie analytique de la chaleur*, published in *Paris, 1822*, while an English translation was made by A. Freeman (Cambridge, 1878; reprinted New York 1955). A second French edition appeared in 1888 as Vol. I, of *Oeuvres de Fourier*, and Gaston Darboux, ed. Vol. II, containing the majority of the rest of Fourier's published works, appeared in 1890.

An empirical relationship between the conduction rate in a material and the temperature gradient in the direction of energy flow was first formulated by Fourier in 1822 (see Fourier 1955). He concluded that "the heat flux resulting from thermal conduction is proportional to the magnitude of the temperature gradient and opposite to it in sign". For a unidirectional conduction process, this observation may be expressed by (*Fourier heat transfer assumption*):

$$\boldsymbol{q} = -k \operatorname{grad} \mathrm{T}, \tag{1.32}$$

where T is the temperature and the heat flux $\boldsymbol{q}$ arises solely from *thermal convection*. This is called *Fourier's law*, and is the *second constitutive equation for the Newtonian fluid flow*. The coefficient $k > 0$ of the *heat conduction* (*thermal conductivity*) is a function of at most $\rho$ and $e$.

## 1.4 Some Remarks

As a first *conclusion*, of the above rather lengthy story of the *"fluid dynamics adventure"* from *Newton to Cauchy*, *Euler*, *Navier*, *Poisson*, *Stokes*, and *Fourier*, it seems important to emphasize (*once again!*) that the concepts of *density*, *internal energy*, and *velocity* in *continuous media* are *defined phenomenologically without* any reference to *molecular structure*.

The *pressure* and *temperature are assumed to be defined for non-equilibrium states* by their *dependence on density and internal energy as observed in the equilibrium state*! The *stress tensor* and *heat (flux) conduction vector* are *linear isotropic functions of the velocity gradient and temperature gradient, respectively!*

Their introduction thus involves *three transport* parameters $\lambda$, $\mu$, and $k$, which determine the *non-equilibrium* behaviour of a *Newtonian fluid.*

To complete the Euler, Navier, and *NSF* equations given above, one has to give the *specific formulas which relate pressure* p, *temperature* T, and transport parameters $\lambda$, $\mu$, and $k$ to $\rho$ and $e$. Because of the special assumption that p and T obey the equilibrium laws, *it is not necessary to take* $\rho$ and *e as basic state variables*. In the highly instructive Course Notes (written in *1953*) by *Lagestrom—but published only in 1964* [11], the reader can find many cases of such specific formulas (see p. 36). As an example, one may prove, using thermodynamics, that $e$ is a function of T alone.

We note also that it is possible to introduce two *free energy formulations*: *Helmholtz* (when density $\rho$ and temperature T are independent variables) or *Gibbs* (when density p and temperature T are independent variables), and this is very well expounded in the recent paper by Betchtel et al. [12]. Hence, when the *density* ρ and *temperature* T are *independent variables*, the *pressure* p and *entropy S* are given by

$$p = \rho \partial \psi / \partial \rho, \; S = -\partial \psi / \partial T, \tag{1.33a}$$

where the *Helmholtz* free energy

$$\psi \, (\rho, \, T) = e(\rho, S) - TS \tag{1.33b}$$

is the Legendre transform of the internal energy e(ρ, $S$) with respect to the thermal variable, from $S$ to T. On the other hand, when the *pressure* and *temperature* are independent variables in the formulation, the specific volume 1/ρ and $S$ are given by

$$(1/\rho) = \partial \phi / \partial p, \; S = -\partial \phi / \partial T, \tag{1.34a}$$

where the *Gibbs* free energy

$$\phi(p, \, T) = \psi(\rho, \, T) + \, p/\rho \tag{1.34b}$$

is the Legendre transform of the Helmholtz free energy ψ (ρ, T), transforming the mechanical variable from the specific volume 1/ρ to the pressure p. In [12], the reader can find the two corresponding formulations of the NSF equations with T as independent variable, for the five primitive scalar unknowns $u_1$, $u_2$, $u_3$, ρ, T, which are in general functions of space $x_1$, $x_2$, $x_3$, and time t, with ψ (ρ, T), $\mu$(T), $\lambda$(T) specified functions and $k$ a specified constant.

A *Newtonian fluid*—so named to commemorate *Isaac Newton's pioneering treatise of 1687* [1], cited in "Introduction"—is defined to be a fluid whose shear stress is linearly proportional to the velocity gradient in the direction perpendicular to the plane of shear. This definition is valid regardless of the forces acting on a fluid, provided that it *continues to flow*. For example, water is a Newtonian fluid, because it continues to display fluid properties no matter how much it is stirred or mixed. A slightly less rigorous definition is that the *drag* of a small object being moved slowly through the fluid is proportional to the force applied to the object. Important fluids, like water as well as most gases, behave to good approximation as a Newtonian fluid under normal conditions on Earth.

By contrast, *stirring* a *non-Newtonian fluid* can leave a "hole" behind! This will gradually fill up over time. Such behaviour is observed in materials such as pudding, oobleck, or sand (although sand isn't strictly a fluid). Alternatively, stirring a non-Newtonian fluid can cause the viscosity to decrease, so the fluid appears "thinner" (this is seen in non-drip paints).

There are many types of non-Newtonian fluids, since they are defined to be anything that fails to obey a particular property—for example, most fluids with long molecular chains can behave in a non-Newtonian manner. In some applications, another broad and rather rough division is made, distinguishing *ideal* and *non-ideal* fluids. An *ideal fluid is non-viscous* (by definition) and offers no resistance whatsoever to a shearing force.

An *ideal fluid could not really exist* (!), but in some calculations, the assumption is justifiable. One example of this is the *fluid flow far from solid surfaces*. In many cases the *viscous effects are concentrated near solid boundaries* (such as in boundary layers), while in regions of the flow field far *away from* the boundaries, *viscous effects can be neglected* and the fluid there is treated as if it were *inviscid* (*ideal flow*). When the viscosity is neglected, the term containing the *shear stress tensor* in the NSF equation for **u** *vanishes*. The equation reduced in this form is just called the Euler equation.

Since the basic (in the more general case, the NSF) equations are known, the question concerning in particular the *Navier, viscous, and homogeneous incompressible cases* is: *How much of a theoretical handle do we have on the Navier equations?* According to *Roberta Benzi* and *Uriel Frish* (2010) in *Scholarpedia*, 5(3): 3439, the *short* answer is *very little*! We cannot, for example, show that the solutions of the Navier equations with nice smooth initial conditions stay nice, smooth, and unique for all times, at *least not in 3D!* But in 2D, *yes, we can!*

It was even speculated by *Jean Leray* [8, 9], cited in "Introduction" in the 1930s that the random character of *turbulence* originates from non-uniqueness of the solutions to the Navier equations! Nowadays, we know enough about how chaos can appear in *deterministic dynamical systems* that there is no need to resort to non-uniqueness to explain turbulence. In our *essay*, Zeytounian, 1989 [13], the reader can read about *some phenomenological aspects of the transition to turbulence*.

When we try to tackle random solutions of the Navier equations, we have to face a closure problem: because the equations are quadratically nonlinear, the time rate of change of the correlation functions of the velocity at $n$ different points involves similar correlation functions, but with $n+1$ arguments. An infinite hierarchy of equations is then obtained. The simplest form of closure, introduced by *Kolmogorov*'s student *Milionshikov*, is to arbitrarily close this hierarchy by relating *fourth-order correlation functions* to *second-order ones* as if the velocity had Gaussian statistics. This is, of course, unjustified and leads to problems such as negative values for *energy-like* quantities, which are by definition non-negative. Cures for such ailments can be found, but they are frequently *ad hoc* with no possibility of controlling the errors made with respect to the correct solutions.

*The main difficulty is the absence of a small parameter which would allow one to start a suitable perturbative RAM approach.* The *smallness of the viscosity*—or equivalently a *large value for the Reynolds number*—has been of no use so far, because very little is understood about the *Euler equations,* that is, the Navier equations with the viscosity set to zero. The most fruitful theoretical approaches have been based on *scaling arguments, hence essentially on dimensional analysis, which is the basis of our RAM approach.*

Another rather fruitful kind of approach is through the use of "*toy models*": having identified certain properties of the basic equations which are believed to play a key role in the behavior of turbulence (for example *invariance* and *conservation properties*), one tries to find simpler models sharing those properties and which can be solved analytically, or at least for which numerical solutions are much simpler than for the full 3D fluid dynamics equations. Precise definitions of such models would take up too much space, but we can give an idea of what has been achieved with the use of some of these toy models. For instance, the *1D Burgers equation* (see, for instance, Sect. 8.9) gives a concrete example of how energy dissipation can have a finite non-vanishing limit when the viscosity tends to zero, in spite of the fact that the *inviscid equation formally conserves energy*. As pointed out by *Saffman*, it also shows what can go wrong with naive application of dimensional arguments.

The random coupling model of *Kraichnan* starts with *N independent replicas of the random Navier equations* and then couples them artificially by random *Gaussian coefficients,* chosen in such a way as to preserve most invariance and conservation laws. In the limit $N \to \infty$, closed equations, called the *direct interaction* (*DIA*) equations, are obtained for suitable statistical quantities (including the two-time and two-point velocity *correlation functions*)—but the solutions are *not compatible* with the *Kolmogorov 1941 theory*. This happens, not because scale invariance is broken, but because the model does not preserve a certain form of *Galilean invariance?* Fortunately, this shortcoming can be repaired by resorting to a kind of *Lagrangian description* (*Kraichnan*) or by making the coupling coefficients scale-dependent to obtain the *eddy damped quasi-normal Markovian* (EDQNM) model (of *Orszag*).

"*Shell models*" start from the Navier equation written in spatial Fourier space and replace all the *Fourier modes* in the shell with wavenumbers between $2n$ and $2n+1$ by just a few degrees of freedom, typically one complex number. The interactions between these "*shell amplitudes*" are of course chosen again to preserve as many features as possible of the original equations. Some shell models, such as *GOY* or *SABRA*, are known to display the same anomalous scaling as the full equations. Unfortunately, little theoretical progress has been achieved and the main advantage of the shell models remains their ability to run very high Reynolds numbers using just a workstation. Much more drastic simplifications of the true dynamics are involved in the multiplicative random model (of *A. M. Yaglom*—who was, with *Lev Gutman,* my examiner, in January 1957 at IFA, in Moscow, before my working years with *I. A. Kibel in the Dynamical Meteorology Department of Guidro-Meteo Center*!). Here the amplitudes in shell $n+1$ are just obtained by multiplying the amplitude in shell $n$ by a random variable with a suitable distribution (independent identically distributed random variables are assumed for different $n$). *Correlation functions* can then be calculated explicitly; they display anomalous scaling and *multifractality* through a mechanism involving large deviations (*Varadhan*). Finally, purely qualitative models can be made, such as the *Swift–Richardson flea-eddy model,* which are stimulating for the imagination.

As stated before, *turbulent flows abound everywhere around us*. However, high-Reynolds-number flows displaying good scaling properties of the kind discussed require very large scales, as is the case in the natural *atmospheric* and *astrophysical environments*. The former keeps changing (with the *weather*) and the latter is not so easily accessed. Large-scale facilities, such as major wind tunnels, are very expensive and it is difficult to have a fundamental experiment running there for a duration of several days to several weeks (as is needed, e.g., to accumulate good *statistical data* on rare violent events) when they are competing with important industrial applications such as the *testing of new designs* for cars and *airplanes*. During the last ten years, a new technique has been developed involving *low temperature Helium*.

Concerning the *turbulent boundary layers*, the reader can find a *rather complicated asymptotic description* by *Deriat* and *Guiraud*, in *1986*—see [10, pp. 109–140], cited in "Introduction".

In the survey article by *Birkhoff* [4], cited in "Introduction", the reader can find a short and well organized account of *analytical fluid dynamics*"—and in fact, one cannot understand *numerical fluid dynamics* without knowing a good deal about analytical fluid dynamics! For instance, Euler stated optimistically in 1755 that "*the whole theory of the motion of fluids has just been reduced to the solution of analytic formulas*". Surprisingly, it was not until 1843 that Stokes determined the *potential* (*irrotational* and *incompressible*) flow around a moving sphere, and he showed that the pressure p is determined (in steady flow) by *Bernoulli's formula*

$$p + (½)\rho_0|\mathbf{u}|^2 + G(\mathbf{x}) = \text{const.},\ \mathbf{g}(\boldsymbol{x}) \quad = \quad -\nabla G(\boldsymbol{x}), \tag{1.35}$$

where $\mathbf{g}(\boldsymbol{x})$ is the gravity. This is valid along any streamline in a fluid of constant density $\rho_0$, in any "steady flow" with time-independent velocity $\mathbf{u}(\boldsymbol{x})$, and $p$ is *symmetric fore and aft*! Stokes' solution confirmed the truth of an *assertion made by* d'*Alembert* 75 years earlier, that *a solid moving through an ideal fluid, as defined by the Euler equations, would encounter no resistance at all! Symmetrical solids should encounter no lift or drag in steady translation. Stokes' profound and lucid essays inaugurated 75 years of great progress in analytical fluid dynamics. This progress is very well documented in Lamb's "Hydrodynamics", whose first edition was published in 1879 and whose sixth edition appeared in 1932.*

But by far the most important progress made in integrating the Navier equation stems from another asymptotic approximation proposed by Prandtl in 1904 and treated by him 30 years later. Prandtl suggested first in 1904 that, for $\mathrm{Re} \gg 1$ (we observe that the *Stokes flow around a sphere is only valid asymptotically, as* $\mathrm{Re}\downarrow 0$), the effect of viscosity when the fluid adheres to the body wall would be confined to a *thin boundary layer (BL)* near the solid wall, in which a rapid transition takes place from near-potential flow to zero velocity near-wake flow.

In this boundary layer, one can *greatly simplify* the Navier equation (1.26a), for instance to (when written with local (inner) boundary-layer functions and primed coordinates):

$$
\begin{aligned}
&\partial u'\partial t + u'\partial u'/\partial u' \; + v'\partial u'/\partial y' + (1/\rho^\circ)\partial p'/\partial x' \; = \; (1/\mathrm{Re})\partial^2 u'/\partial y'^2,\\
&\partial u'/\partial u' \; + \partial v'/\partial y' \; = \; 0,\\
&\partial p'/\partial y' \; = \; 0,
\end{aligned}
\tag{1.36}
$$

but for a *uniform flow past a flat plate*—the pressure $p'$ being presumably determined by the (outer) non-viscous potential flow, just outside of the boundary layer, where the Euler equations are valid, *via a matching process*, as discussed in Sect. 3.3.

Unfortunately this so-called "*non-interacting coupling of Euler and Prandtl flows* for $\mathrm{Re} \gg 1$" *does not work when the non-viscous (inviscid) Eulerian free stream line* leaves *the neighborhood of the body*. At this *"detachment point" a "separation" appears, and* in *1948 Golstein* showed that the solution of the Prandtl BL equations *could not be continued downstream of separation*. Incidentally, it *does not seem possible to confine* the effects of this *singularity* to the *neighborhood of the detachment/separation point*! In 1970, in *J Fluid Mechanics*, vol. 44, pp. 347–364, *Stewartson* wrote: "*Is the singularity at separation removable*?" The answer was provided *independently* by *Neiland (1969)* and *Stewartson and Williams (1969), using a new scheme for the flow, the so-called "triple-deck model", already mentioned above,* which has proved to be extraordinary fruitful, both practically and theoretically, thanks to various papers by many authors: *Lighthill (1953), Brown and Stewartson (1969), Messiter (1970), Bouthier (1972), Sychev (1972), FT Smith (1973), Guiraud (1974), Sykes (1978), Daniels (1978), Guiraud and Zeytounian (1979), Ruban and Sychev (1979), FT Smith, Brighton, Jackson and Hunt (1981), Meyer (1983), Stuart (1986), Nayfeh (1991), Lagree (1992), Degani, Walker and FT Smith (1998)*. The reader can find the references of the above-mentioned authors in pp. 542–545 of [7], cited in "Introduction".

During the first half of the last century, airplanes progressed from an unrealized dream to a major factor in human transportation and military power. As a by-product, aerodynamics became the most exciting research frontier of fluid mechanics. Among the many new mathematical models of airflows that were constructed during this period, five stand out. Only 50 years later Prandtl's, fruitful boundary layer idea—using *asymptotics*, initiated in California by *Lagerstrom, Cole, and Kaplun* (see the book by *Van Dyke* [16], cited in "Introduction")—

*inaugurated* an amazing new period of great progress in theoretical fluid dynamics. Personally, after my return from Moscow to Paris and thanks to a careful study of a paper by Germain and Guiraud [14], *ONERA* (France), entitled "*Asymptotic theory of jump conditions and structure in gas dynamics for high Reynolds number*", I succeeded during the years 1969–1974 in justifying the bold *assertion by Boussinesq (1903)* laid down in *Boussinesq's model equations*; and *in* Sect. 2.6 of the present book, the reader can find an asymptotic *derivation of these unsteady Boussinesq equations* from the Euler compressible equations. Indeed, my first conception of *meteorology as a discipline of fluid dynamics and a privileged area for the RAM approach* was a direct and fruitful consequence of an invitation from the Polish Academy of Sciences to write a survey lecture for the *XIIth Symposium on Advanced Problems and Methods in Fluid Dynamics,* in Bialowieza, in 1975, published in 1976 [21], cited in "Introduction".

This vision of atmospheric motions from a Newtonian fluid dynamics point of view leads one to consider a more complicated but very rich system of dimensionless equations with a large family (see Sect. 3.4) of reduced dimensionless parameters. As a consequence, the RAM approach can be used to derive many rational reduced approximate working models, which can be very useful, not only in the framework of weather forecasting, but also in the study of *terrestrial/environmental* fluid dynamics! But it seems (see pp. 258–259 in the *companion 2012* book) that the French weather office *Meteo France prefers a different, rather questionable approach*! Despite this, I think that viewing atmospheric motions as *terrestrial fluid dynamics* raises numerous interesting and specific problems of a challenging nature regarding *our environment.* In Sects. 8.6 and 8.7, the reader can find *two particularly interesting meso-scale weather problems.* Sections 3.4 and 6.3 aim to classify *various working non-ad hoc model problems for atmospheric motions, derived from the full unsteady NSF atmospheric equations using the RAM approach.*

Clearly, Chap. 1 gives only a very short and partial introduction to the fluid dynamics "*adventure*" from Newton to Stokes! But in the book by *Tokaty, 1994,* the curious reader is referred to the unabridged, corrected edition (*by Dover publications)* of the book entitled "*A History and Philosophy of Fluid Mechanics*", published by Foulis, Henley on Thames in 1971, which is especially interesting for its illuminating comparison between the development of fluid mechanics in the former Soviet Union and that in the West.

The above book by *Tokaty* concludes with studies of *transonic compressibility effects on lift.* When an aircraft goes through this *transonic region (a whole range of Mach numbers, say, from 0.82 to 1.2, because transonic flows are mixed flows, and include both low supersonic and high subsonic regimes), unexpected changes can occur in the trim. Because of the appearance of a shock wave (or waves) on the upper surface of the wing, the point of action of the resultant lift may be suddenly displaced, thereby disturbing the relative locations of the lift and weight forces.* Then, in *supersonic flow, when there is a shock wave before a body, this leads to a certain decrease in the drag in the body.*

*An important and interesting feature of Mach waves is that, when they are waves of expansion (rarefaction), they can never come together and reinforce each other, but always spread apart like a fan.* On *the other hand, when they are waves of compression, they frequently do meet and reinforce each other. In such regions of reinforcement, we have a large number of weak compression waves adding up to form a front, across which a definite and almost discontinuous compression takes place—a shock wave.* This theory was developed first by *Th. Meyer (1908) and later on by J. Ackeret, L. Prandtl, and A. Busemann.*

*H.S. Tsien (1946)* is, really the disoverer of the hypersonic gas dynamics (very high Mach number m$\gg$1). This is closely associated with rockets and space vehicles, which have to fly in different physical conditions, from the normal atmospheric continuum to the extremely rarefied fringe layer of the atmosphere. His theory provoked an almost immediate reponse from several other "fathers" and thereby paved the way to broader investigations; in the USA by *Hayes (1947)* and *Hayes and Probstein (1966)*, in France by J.P. Guiraud *(*1963/1965*)* [15], and in the former Soviet Union by *G.G. Chernyi* (his book was translated into English and published by *Probstein in 1961*).

Note that *continuum gas dynamics first had to be modified, then abandoned (!),* because at *very high altitudes*, the atmosphere becomes so rarefied that it no longer behaves like a continuous fluid. The basic molecular character of air then gives rise to important modifications in aerodynamic and heat transfer phenomena., The reader can find a short discussion in *Tokaty (1994), pp. 225–230.* As a conclusion Tokaty wrote:

> It should not be thought, however, that hypersonic gas dynamics is the science of rocketry and space technology alone—no, it has a much wider value—we shall have new types of space vehicles, and we shall in future have also hypersonic aircraft. Who would even have dreamt only (say) 25 to 30 years ago that there would be supersonic aircraft like the Concorde and Tu-144?

But today they are with us, and there can be no doubt whatsoever that *hypersonic air transport is also on its way. Hence, hypersonic gas dynamics is no longer merely a science of rocketry and re-entry vehicles.*

In the book by Theodore von Karman [16], the reader can find *selected topics on aerodynamics, taking into account the historical development.* Authoritative and engaging, this popular history traces the science of aerodynamics from the age of Newton right through the mid-twentieth century!

We observe that *in addition to subsonic, transonic, supersonic, and hypersonic gas dynamics, there is also hyposonic flow theory,* and our pioneering book Zeytounian [24], cited in "Introduction", is devoted entirely to *low Mach number fluid (hyposonic) flow*, when the dimensionless Mach number M is small relative to unity. The reader can find various *geophysical* and *environmental* fluid flow problems which are also relevant to engineers because of their connection with classical *aerodynamics* problems, *nonlinear acoustics,* and *combustion* theory.

In Sect. 8.5, the reader can also find an interesting example related to a *generalization* of the *Blasius classical (incompressible) viscous problem to* a *slightly compressible viscous fluid flow*.

As a complementary topic related to the above account, the curious reader is invited to read our review paper Zeytounian [2], which contains a *historical survey of some mathematical aspects of Newtonian fluid flows*.

## References


[1] Segel LA (1987) Mathematics applied to continuum mechanics. Dover

[2] Zeytounian RKh (1991) A historical survey of some mathematical aspects of Newtonian fluid flows. Appl Math Rev 54(6):525–562

[3] Yih C-S (1980) Stratified flows. Academic Press, New York, NY

[4] Zeytounian RKh (1995) Nonlinear long waves on water and solitons. Physics-Uspekhi 38(12):1333–1381

[5] Newell AC (1985) Solitons in mathematics and physics. SIAM, Philadelphia

[6] Stewartson K (1981) D'Alembert paradox. SIAM Rev 23:308–345

[7] Guiraud JP (1995) Going on with asymptotics. In: Bois P-A et al (eds) Lecture notes in physics—LNP 442. Springer, Heidelberg

[8] Zeytounian RKh (1974) Notes sur les Ecoulements Rotationnels de Fluides Parfaits. Lecture notes in physics, vol 27. Springer, Heidelberg

[9] Zeytounian RKh (1966) Three-dimensional rotational motion in a compressible fluid. Izv Atmos Oceanic Phys 2:61–64

[10] Marchioro C, Puvirenti M (1994) Mathematical theory of incompressible non-viscous fluids. AMS 96. Springer, New York, NY

[11] Lagerstrom PA (1964) Laminar flow theory. In: Moore FK (ed) Theory of laminar flows—section B. Princeton University Press, pp 20–285

[12] Betchtel SE et al (2004) Investigation of simplified expansion. Phys Fluids 16, 11-November

[13] Zeytounian RKh (1989) An essay on the transition to turbulence. Arch Mech 41(2/3):383–418

[14] Germain P, Guiraud JP (1966) J Math Pures Appl 45:311–358

[15] Guiraud JP (1963/1965) Topics in hypersonic flow theory. Isd. MIR, Moscou (1965) [in Russian—translation from the English of lectures given in: Department of Aeronautics and Astronautics, Stanford University, and edited as SUDAER, TP N0 521, 1963, Stanford, CA]

[16] von Karman Th (1954) Selected topics in the light of their development. Reprint of the Cornell University Press, New York

# Part I
# Classical Analytical-Asymptotics Newtonian NSF Fluid Dynamics

# Chapter 2
# Formulation of Some *NSF* Unsteady Initial-Boundary Value Problems

In the present chapter, using the results of Sects. 1.1–1.3 of Chap. 1, the reader can find *first* a mathematical formulation of a typical, initial-boundary value, unsteady NSF problem in the case of a *thermally* (calorically) *perfect gas*. Then, a more specific NSF problem of this kind is formulated for an *expansible/dilatable* liquid under the influence of temperature, with a specific equation of state; this is use to show the emergence of the working models for classical *Bénard thermal convection, heated from below*. The corresponding NSF problems are also formulated for *nonlinear acoustics* and *atmospheric motions*.

We also mention the case of so-called *Navier-Stokes* (NS) *isentropic*, *viscous*, compressible fluid flow, *a physically inconsistent case,* mainly considered by mathematicians, in their *pure rigorous investigations using abstract nonlinear functional analysis*!

Concerning *large horizontal scale atmospheric motions*, the *gravity term* means that one must consider the *influence of the Coriolis force* in the momentum equation for the velocity vector as observed in the *rotating earth frame*, and one must also employ *spherical coordinates*. In this case we are confronted with a *very stiff, complicated, starting system of dimensionless equations,* but this system allows one to use the RAM approach to derive various reduced working models.

It is important to note that *the equations are not sufficient for the applications of the RAM approach and deconstruction analysis* to the NSF system of equations. One must also formulate *physically convenient initial and boundary conditions.* Indeed, the equations given below govern the flow of a fluid—they are the same equations whether the flow is, for example, over an Airbus A380 (cruising speed 912 km/h) through a subsonic wind tunnel or past a windmill. The flow fields are quite different for these cases, even though the governing equations are the same.

Why? Where does the difference enter? The answer is through the boundary conditions, which are quite different for these two examples. The boundary conditions, and sometimes the initial conditions (with given data), dictate the particular solutions to be obtained from the governing equations (see Sect. 2.5). The above remark seems trivial, but unfortunately, in mathematically rigorous investigations,



R.Kh. Zeytounian, *Challenges in Fluid Dynamics*,
DOI 10.1007/978-3-319-31619-2_2

the role of the associated conditions is often not given serious consideration by mathematicians who have the strange opinion of a futility concerning these conditions! Some complementary remarks are made in Sect. 2.7.

Concerning the RAM approach, the reader can find the "*mathematics*" of this approach in Chap. 6 of our *companion 2012* book, but in Chap. 3 of the present book, I have decided, as explained in the preface, to provide some complementary enlightenment!

## 2.1 The Case of a Thermally Perfect Gas: Typical NSF Equations

For a *thermally perfect gas* (identified as *calorically perfect,* for simplicity), we first write the following equation of state for the specific internal energy e:

$$e = Cv\,T, \tag{2.1}$$

where Cv is the specific heat for a specific volume. With the various results given in Chap. 1, we are now in a position to write the following full unsteady NSF equations for a thermally perfect gas, where we assume that the kinematic viscosity $\nu = \mu/\rho$ and also the coefficient of thermal conductivity k are both constant coefficients. Hence, we have the following NSF system of equations governing the fluid flow for the four unknown functions $\mathbf{u}$, ρ, p, and T:

$$\rho D\mathbf{u}/Dt = \rho \boldsymbol{f} - \nabla p + \{\nabla[\lambda(\nabla.\mathbf{u})] + \nabla.[2\mu\underline{\mathbf{D}}]\}, \tag{2.2a}$$

$$D\rho/Dt + \rho\,\mathrm{div}\mathbf{u} = 0, \tag{2.2b}$$

$$\rho Cv\, DT/Dt + p\,\mathrm{div}\mathbf{u} = k\nabla^2 T + \Phi, \tag{2.2c}$$

$$p = R\rho T, \tag{2.2d}$$

where R = Cp − Cv is the gas constant and Cp the specific heat for p constant and

$$\Phi = \lambda(\nabla.\mathbf{u})^2 + 2\mu\underline{\mathbf{D}} : \underline{\mathbf{D}}. \tag{2.2e}$$

The ratio γ = Cp/Cv may be taken as unity for a pressure near normal values, and λ is the second viscosity coefficient (often assumed *constant*).

The above system of equations (2.2a–d), with (2.2e), is just the full unsteady/evolution (in time), typical NSF classical system of *aerodynamics* equations (often written with $\boldsymbol{f} = 0$).

For *atmospheric* motions, when dry atmospheric air is treated simply as a perfect gas, one must consider *gravity* $\mathbf{g} = -g\mathbf{k}$ as an external force in place of $\boldsymbol{f}$ in (2.2a), with the unit vector **k** in the vertical direction opposite to the gravity vector **g**. For large horizontal scale atmospheric motions, when the NSF equations are written in a system of spherical coordinates (see Sect. 3.4), in a coordinate frame rotating with the earth, one must use an equation for the *relative velocity* taking into account the *Coriolis force,* in place of equation (2.2a). In a such a case, as done in Sect. 3.4, the RAM approach allows us to derive a multiplicity of working non-*ad hoc* models simulating various interesting and useful atmospheric motions. These are sketched on Fig. 6.3 of Sect. 6.3.

On the other hand, if we take into account (2.1), with (2.2e), and the above formulas for R and $\gamma$, then in place of equation (2.2d) for the pressure p, we can use the following equation of state:

$$p = \rho(\gamma - 1)e, \ \gamma = \frac{C_p}{C_v}. \tag{2.2f}$$

We also observe that a rather more complicated formulation of the NSF equations for a compressible, viscous, and heat conducting Newtonian fluid flow is possible when, instead of the equation of state, we introduce the *Helmholtz* or *Gibbs free energy* formulations in thermodynamics (mentioned in Sect. 1.4).

## 2.2 The Case of an Expansible Liquid

The equation of state for various liquids is usually taken to give the density $\rho$ as a function of T and p:

$$\rho = \rho(T, p). \tag{2.3a}$$

In Rayleigh <1>, a pioneering paper entitled "*On convection currents in a horizontal layer of fluid, when the higher temperature is on the under side* », instead of (2.3a), it is assumed that

$$\rho \approx \rho(T), \ \text{with} - (1/\rho)\, d\rho/dT = \alpha(T), \tag{2.3b}$$

where $\alpha$ (T) is the volume/thermal expansion coefficient, and measurements show rather *smaller values* for the above coefficient $\alpha(T)$ for liquids than the value 1/T appropriate to a thermally perfect gas. In reality, rather than the state equation (2.3a), it is usually assumed (see for instance [25], cited in "Introduction") that the expansible liquid can be described by the following *approximate* (*truncated*) law:

$$\rho = \rho_d[1 - a_d(T - T_d) + \chi_A(p - p_A)], \tag{2.4a}$$

where $\rho_d$, $T_d$, and $p_A$ are some constant values for the density, temperature, and pressure, and $\chi_A$ , $a_d$ are respectively the so-called *isothermal compressibility* coefficient:

$$\chi_A = (1/\rho_d)\,[\partial\rho/\partial p]_A, \tag{2.4b}$$

and the *coefficient of thermal expansion at* $\rho_d$ and $T_d$:

$$\alpha_d = -(1/\rho_d)\,[\partial\rho/\partial T]_d. \tag{2.4c}$$

In his 1916 paper <1>, Rayleigh wrote: "*Bénard worked with a very thin layer, in his 1900/1901 experiments* [5], cited in "Introduction"." In the highly relevant monograph by *Chandrasekhar (1981),* the author explains that *a depth of only about 1 mm is considered on a level metallic plate; this layer was usually free, and being in contact with the air was at a lower temperature*—various liquids were employed, and the *layer rapidly resolved itself into a number of cells.*

More precisely, Bénard found that when the temperature of the lower surface was *gradually increased,* at a *certain instant* (*bifurcation—see* Chap. 10 *of* [14], cited in "Introduction"), the *layer became reticulated* and revealed its *dissection into cells.* He further noticed that there were motions *inside the cells*: of *ascension* at the centre, and of *descension* at the boundaries with the adjoining cells.

With the specific enthalpy h, the energy equation is

$$\rho Dh/Dt - Dp/Dt = k\nabla^2 T + \Phi. \tag{2.5}$$

In our 2009 monograph [25], cited in "Introduction", devoted to *convection in fluids*, the reader can find a *unified RAM approach* to the three main convection cases: shallow *thermal* (Rayleigh-Bénard), *deep thermal* (à la Zeytounian), and *thermocapillary* (Bénard-Marangoni) convections.

Concerning the equation for the (absolute) temperature T associated with the above full equation of state (2.3a) for the liquids, we have:

$$((C_p)/\gamma)\rho DT/Dt = (\alpha_d/\chi_A)T(\nabla.\mathbf{u}) = k\nabla^2 T + \Phi. \tag{2.6}$$

Hence, the starting NSF equations for an expansible liquid are the two Navier-Stokes dynamic equations (2.2a) and (2.2b), with the equation of state (2.3a) and equation (2.6) for the temperature, where $\Phi$ is given by (2.2e).

## 2.3 Navier-Stokes (NS) Barotropic Compressible Equations

In the case of *NS equations, only a reduced system of two equations is considered,* rather than the full NSF equations: the dynamic equation (2.2a) for $\mathbf{u}$, where $\lambda$ and $\mu$ are both assumed constant and designated by $\lambda_O$ and $\mu_O$, with continuity equation (2.2b) for $\rho$, and also a simplified specifying equation between p and $\rho$, for a so-called elastic/barotropic compressible and viscous fluid. We thus have:

$$\rho \mathrm{D}\mathbf{u}/\mathrm{Dt} = \rho \mathbf{f} - \nabla p + \mu_O \nabla^2 \mathbf{u} + (\lambda_O + \mu_O)\nabla[\nabla.\mathbf{u}], \tag{2.7a}$$

$$\mathrm{D}\rho/\mathrm{Dt} + \rho \nabla.\mathbf{u} = 0, \tag{2.7b}$$

$$\mathrm{p} = \mathrm{P}(\rho), \tag{2.7c}$$

which forms a simplified closed system of three equations for **u**, ρ, and p. This is the so-called Navier-Stokes compressible, but isentropic, system of equations, a system often used by "pure mathematicians" in mathematical fluid dynamics, as in Lions (1998)!

The above NS system of three equations (2.7a)–(2.7c) is a *correction,* disregarding viscosity, of the Euler system (1.20b,c) given in Sect. 1.2, with the above specific equation (2.7c) in the place of the equation of state $\mathrm{p} = \exp(\mathrm{S/Cv})\rho^{\gamma}$, and without the equation DS/Dt = 0 for the specific entropy. These compressible non-viscous equations for an *elastic fluid* in an adiabatic/barotropic flow, first considered by Euler in 1755, constitute, *in contrast, a very consistent non-viscous fluid flow model,* with many interesting and varied applications. Unfortunately, a barotropic and compressible fluid flow, when assumed viscous, as in the above NS system of equations (2.7a)–(2.7c), does not have, on the contrary, any physical reality, and obviously the various rigorous results obtained by mathematicians using these NS equations are of little interest!

Viscosity always generates entropy—baroclinity—the typical case being atmospheric motion for dry air, treated as a trivariate thermally perfect gas, as in Sects. 2.6 and 2.4. This was noticed by Leray (1994) in relation to rigorous mathematical abstract results derived for low Mach number asymptotics by P.L. Lions. But despite Leray's critical observation, 5 years later, in a survey (!) by Desjardins and Lin (1999), this isentropic viscous, physically inappropriate model was discussed once again (!), while the critical paper by *Leray* (1994) was curiously ignored.

## 2.4 The Case of Nonlinear Acoustics

As a *dominant acoustic* system of equations *à la NSF* for the velocity vector **u** and the thermodynamic perturbations, π, ω, and θ, such that

$$\mathrm{p} = 1 + \mathrm{M}\pi, \quad \rho = 1 + \mathrm{M}\omega, \quad \mathrm{T} = 1 + \mathrm{M}\theta, \tag{2.8}$$

and *neglecting* the terms proportional to $O(M^2)$—see, for instance, pp. 17–18 of my book 2006 [24], cited in "Introduction"—we obtain the following system of equations, assuming that the bulk viscosity is zero, i.e., $\lambda + (2/3)\mu = 0$ according to the Stokes relation:

$$\partial\omega/\partial t + \nabla.\mathbf{u} = -\mathrm{M}\,[\nabla.(\omega\mathbf{u})], \tag{2.9a}$$

$$\partial\mathbf{u}/\partial t + (1/\gamma)\nabla\pi = (1/\mathrm{Re}_{ac})\{\nabla^2\mathbf{u} + (1/3)\nabla(\nabla.\mathbf{u})\} - \mathrm{M}\,[\omega\partial\mathbf{u}/\partial t + (\mathbf{u}.\nabla)\mathbf{u}], \tag{2.9b}$$

$$\partial\theta/\partial t + (\gamma-1)\nabla.\mathbf{u} = (\gamma/\mathrm{PrRe}_{ac})\nabla^2\theta + \gamma(\gamma-1)\,(\mathrm{M}/\mathrm{Re}_{ac})\{(1/2)\underline{\mathbf{D}}:\underline{\mathbf{D}} + (1/3)\,(\nabla.\mathbf{u})^2\} - \mathrm{M}\,[\omega\partial\theta/\partial t + (\mathbf{u}.\nabla)\theta + (\gamma-1)\pi\nabla.\mathbf{u}], \tag{2.9c}$$

$$\pi = -(\theta + \omega) + \mathrm{M}\theta\omega. \tag{2.9d}$$

The above "dominant" equations (2.9a)–(2.9d) are the main starting point for the derivation of various model equations in *nonlinear acoustics*, considered as a branch of fluid dynamics by Crighton [6], cited in "Introduction", Lighthill, *1954*, and also Coulouvrat <2>.

For the frequencies and media commonly used in nonlinear acoustics, the acoustic Reynolds number $\mathrm{Re}_{ac}$ in the above equations (2.9b) and (2.9c) is always very large compared to unity. This means that the (continuous) medium is weakly dissipative at the frequency chosen in the most common experimental situations, and the above system of equations (2.9a)–(2.9d) is quite accurate enough. As a consequence, the above three unsteady equations turn out to be perturbation system of equations with two small parameters M and $1/\mathrm{Re}_{ac}$. But neglecting the terms proportional to $O(M^2)$, the above system makes sense only if

$$\mathrm{M}^2 << 1/\mathrm{Re}_{ac}. \tag{2.10}$$

In Sect. 8.9, the reader can find a very short derivation of the famous KZK parabolic single equation model for low acoustic Mach number incorporating *nonlinearity, dissipation,* and *diffraction* effects. This KZK amplitude equation is in fact a *generalised form* of the well-known *one-dimensional Burgers equation,* derived in 1948, illustrating the *theory of turbulence,* when the parameters linked with the relative orders of magnitude of diffraction and nonlinearity in the KZK equation *are zero.*

More particularly, concerning the *asymptotic derivation of the KZK model* in Sect. 8.9, the main idea is basically that the *3D acoustic field is locally plane, so that the nonlinear wave propagates in the same way as a linear plane wave over a few wavelengths.* In this case, the *wave profile or amplitude is significantly altered* only at *large distances* from the source, i.e., *in the far field!*

## 2.5 Initial-Boundary Value Problem for the Typical NSF Equations

The boundary conditions, and sometimes initial conditions, dictate the particular solutions to be obtained from the governing NSF equations formulated above. When we consider a fluid flow governed by the typical NSF system of equations, e.g., by the three evolution equations (2.2a)–(2.2c), with the equation of state (2.2d) for a thermally perfect gas, it is first necessary to have a complete set of initial conditions, with *given data*, $\mathbf{u}^\circ$, $\rho^\circ$, $T^\circ$, for $\mathbf{u}$ , ρ, T:

$$t = 0 : \mathbf{u} = \mathbf{u}^\circ(\mathbf{x}), \rho = \rho^\circ(\mathbf{x}) \text{and} T = T^\circ(\mathbf{x}). \tag{2.11a}$$

The typical case is obviously linked with weather forecasting and the crucial question: what will the weather be like tomorrow and for the next few days! But, these initial conditions (2.11a) are also necessary in various unsteady fluid dynamics problems—see, for instance, the problem considered in Sect. 8.8.

The case when *at time* $t = t_{ac}$ an *accident/disaster* takes place during the unsteady evolution of a fluid flow with time is also a case for which the RAM approach can be useful! A spectacular, but very dramatic example is the explosion of the space shuttle Challenger on its tenth launch, on January 28, 1986. This was indeed a frightful disaster. The main cause was the failure of the aft joint seal in the right solid rocket boosters due to the cold weather, leading to a combustion gas leak through the aft field joint of the right solid rocket motor, initiated at or shortly after ignition, which eventually weakened and/or penetrated the external tank, causing structural break-up and loss of the space shuttle during STS Mission 51-L.

When our RAM approach is applied with the above initial conditions, another intriguing aspect which appears as a consequence of the filtering of short compressible acoustic waves during the transition from compressible to incompressible fluid flow, characterize the singular nature of the unsteady low Mach number limiting process near the initial time. Section 3.4 gives some information concerning this filtering and its relation with matching, a basic concept of the RAM approach!

Concerning the boundary conditions, we need to distinguish the case of a fluid flow *extending to infinity (external flow)* in space-time, for which we set various conditions on $\mathbf{u}$ and T when $|\mathbf{x}| \to \infty$. In the case of a *solid body moving* with a given velocity $\mathbf{u}_w$, in the general case of an *NSF problem*, we set the following no-slip condition (a viscous fluid *grips the wall* Γ):

$$\mathbf{u} = \mathbf{u}_w, \quad \text{on } \Gamma, \text{ when } (|\mathbf{x}|, t) \in \text{body}. \tag{2.11b}$$

On the other hand, in the conductive case, $k > 0$ in (2.2c), a boundary condition has to be imposed on the temperature T. In particular, we can require the following *thermal condition*:

$$T = T_w + \chi\,\Theta, \ \text{on}\,\Gamma, \tag{2.11c}$$

where the scalar $\chi > 0$ is a given constant—a measure—for the wall temperature field $\Theta$ (often *in a bounded region on the wall* $\Gamma$), and $T_w$ is a given constant temperature.

The unsteady character of the *typical NSF initial-boundary value problem,* (2.2a)–(2.2d) *with* (2.11a)–(2.11c), as formulated above, is linked mainly with the change in the wall temperature field $\Theta$ with time t, for some evolution problem. In various applications of this problem, the role of the time t in the wall temperature field $\Theta$ must be made more precise in the above thermal condition (2.11c)! This is precisely the case in the above-mentioned problem linked with the crash of the space shuttle Challenger.

The well-posedness of the above typical initial-boundary value (*I-BV*) *NSF fluid flow problem* follows to some extent from properly formulated initial value conditions (2.11a) and boundary conditions (2.11b) and (2.11c). *Unfortunately, rigorous* and valuable mathematical results concerning the existence and uniqueness of this above *I-BV NSF fluid flow problem, i.e., ones that can actually be applied, seem to me outside the scope of "pure" mathematically rigorous investigations by mathematicians.*

Despite many and varied *rigorous papers* published recently by Elsevier B.V. (editors S. Friedlander and D. Serre) in "*Handbook of Mathematical Fluid Dynamics*" during the years 2002 to 2003, 2005 and 2007, it is *very difficult* to extract any helpful result from among them.

*In the framework of the present book, the initial-boundary value NSF fluid flow problem* (2.2a)–(2.2d) *with* (2.11a)–(2.11c) *formulated above is a typical working NSF problem and is designed for use with the RAM approach, when this is rewritten in dimensionless form, as is the case* Sect. 3.2, *and also in* Sect. 3.4 *for atmospheric motions.* Hence, our main objective is the "*deconstruction*" of the *I-BV NSF fluid flow problems formulated above,* first by *rewriting* in *dimensionless form* and then, thanks to the appearance of various *non-dimensional parameters* (numbers) in this *dimensionless (I-BV) NSF problem,* by *application of the RAM* approach.

In Chap. 6, such a deconstruction is achieved for various useful fluid flows, and in Figs. 6.1–6.4, the resulting family of working models are indicated for $Re \gg 1$, $M \ll 1$, *meteo-fluid dynamics*, and the *Bénard convection* problem, respectively. The RAM approach is also applied systematically to various fluid dynamics problems in Sects. 8.1–8.9.

## 2.6 The Rotating Earth and Its Atmosphere as a Continuum

This rather long section is motivated mainly by my long-standing interest in modelling atmospheric motions and the publication of two books: [22], cited in "Introduction" and <3>, a course <4>, and also various papers, e.g., [21], cited in "Introduction" <5>, <6>, and <7>.

The critical review by *Fred R. Payne* (from the Aerospace Engineering University of Texas) in Appl. Mech Rev. 46(42) B29, 1993, of my 1991 book <3>, published by Springer (Heidelberg) under the title "*Meteorological Fluid Dynamics*", describes our approach to atmospheric flows from the point of view of a fluid dynamicist:

> It is apparent that the author expended a considerable effort to make the material in this book accessible to non-specialists. The author has in mind a sequence of models leading to: "...a complete and consistent rational modeling of atmospheric phenomena...in the future." His goal of a: "return of meteorology to the family of fluid mechanics", is both admirable and essential, somewhat like Prandtl's, in 1904, enabling the reunion of (inviscid) aerodynamics and "practical fluids" (hydraulics) after a century of separation.
>
> Appendix 1 is quite a good survey of matched asymptotic expansions (MAE) for singular perturbation problems prevalent in boundary layers but a new prospective user will need to peruse some cited background books.
>
> I found Chaps. 7–10 of most interest. This book seems more suited to individual researchers seeking entry, or as a tutorial by a teacher with some PhD candidates in the area, rather than as a conventional text for a lecture course. Its many examples will be useful in a variety of settings; it is also suited to self-study by an advanced student.
>
> The teacher will need to provide some bridging material, both computational and physical insights, in the early chapters. Chapters 7–9 are written in a more expository style and are essentially self-contained; these chapters make a good text for a one-semester course and would require minimal amplification. As the book stands, it is a sketch for a first course and rather complete for a second.
>
> This is a stimulating book.
>
> Most first-time readers will likely make copious comments upon the margins. It is a member of a select subset, needful of study by specialists in fluid mechanics, turbulence, atmospheric dynamics and modeling, and finite-dimensional dynamic systems.

I feel that my "*Meteorological Fluid Dynamics*" is a good preparation for reading the more ambitious monograph "*Asymptotic Modeling of Atmospheric Flows*", despite the fact that it was published a little earlier in 1990 [22], cited in "Introduction". Concerning this book, *Huijun Yang* of the University of Chicago wrote the following in the book review of SIAM Rev (33, Dec. 1991, pp. 672–673):

> This Asymptotic Modeling of Atmospheric Flows, is rather a monograph in which the author has set forth what are, for the most part, his own results and this is particularly true of Chaps. 7–13. In the book, the author viewed meteorology as a fluid mechanics discipline.
>
> Therefore, he used singular perturbation methods as his main tools in the entirety of the book...The book consists of the author's more than 25 years work. In the thirty-two references of his own work, fewer than one third were published in English, with the rest in Russian or French.
>
> Throughout the book, the reader can strongly feel the influence of Soviet works on the author. However, the author does have his own boundary layer treatment, and well

posedness and ill posedness of the system are very important problems facing researchers today in atmospheric sciences and other related sciences. The reader will find some valuable information on these issues…

The mathematically consistent treatment of the subject does give this book a unique place on shelves of libraries and offices of researchers…

This book is very different from recent books on the market (i.e., Holton (1979), Gill (1982), Haltiner and Williams (1980), Pedlosky (1987), and Yang (1990)).

I recommend that researchers in atmospheric dynamics and numerical weather prediction read this book to have an alternative view of deriving atmospheric flow models. Researchers in theoretical fluid mechanics might also be interested to see how singular perturbation methods can be used in atmospheric sciences.

The book may be used as supplemental material for courses like numerical weather prediction or atmospheric dynamics.

However, I do not think it is a suitable textbook for a regular class: as the author said in his preface—I am well aware that this book is very personal, one might even say impassioned. Unfortunately, in France, and for instance by Roger Teman, the very no-adequate book of Pedlosky (1987) is systematically used in various papers in journal (Nonlinearity), with J.L. Lions and S.Wang (1991, 1992) devoted to atmospheric motions, which are not any practical interest for the modeling the atmospheric motions!

### *2.6.1 The Rotating Earth*

Concerning *meteo fluid dynamics*, one must first observe that the earth revolves about its axis once every 23 H 56 min and 4 s, or a total of 86,164 s. The frequency of rotation or the angular velocity of the earth is

$$\Omega_O = 2\pi/86164 = 7.292 \times 10^{-5}\mathrm{rad/s}, \tag{2.12a}$$

and the radius of the earth at a latitude of $\varphi = 45°$ is $a_O = 6370.1$ km. The true gravitational acceleration owing to the pull of the earth, on the surface and at a geographic latitude of $\phi = 45°$, is

$$|\mathbf{f}| = 9.\,82357\ m/s^2, \tag{2.12b}$$

and therefore, it will be assumed here that the body force, $\rho f$, in the momentum equation is the true gravitational force, where $\rho$ is the atmospheric density.

To distinguish the experiences of a fixed and a rotating observer, let a subscript a denote quantities referred to an absolute, inertial frame of reference, and a subscript r, quantities referred to a frame rotating with the angular velocity of the earth, $\boldsymbol{\Omega} = \Omega_O\, \mathbf{e}$ relatively to the absolute frame. Let $\mathbf{i}, \mathbf{j}$, and $\mathbf{k}$ denote respectively the unit vectors pointing east, north, and vertically upward. Then,

$$\mathbf{e} = \mathbf{k}\sin\varphi + \mathbf{j}\cos\varphi. \tag{2.12c}$$

If $\boldsymbol{\gamma}_r$ is the relative acceleration and $\boldsymbol{\gamma}_a$ the absolute acceleration, we can write the following formula:

$$\gamma_a = \gamma_r + 2\Omega \times v_r - (\Omega_O)^2 x_\perp, \tag{2.12d}$$

where the subscript $\perp$ denotes the equatorial component of $\mathbf{\Omega} \times (\mathbf{\Omega} \times \mathbf{x}) = \vec{\mathbf{\Omega}} \times (\vec{\mathbf{\Omega}} \times \vec{\mathbf{x}})$. For the gravitational acceleration, we write

$$\mathbf{g} = \mathbf{f} + (\Omega_O)^2 \mathbf{x}_\perp, \text{such that} |\mathbf{g}| = g = |\mathbf{f}| - (\Omega_O)^2 |\mathbf{x}_\perp|, \tag{2.12e}$$

and $g = 9.8066\ \mathrm{m/s^2}$ at $\varphi = 45°$. The Froude number Fr—a measure of the significance of the gravitational acceleration (the force of gravity)—is defined by

$$\mathrm{Fr}_{L0} = U_0/(gL_0), \tag{2.12f}$$

where $|\mathbf{U}| = U_0$ , $\mathbf{U}$ is the characteristic constant reference velocity, and $L_0$ is a characteristic length.

If $\mathbf{u}$ denotes the relative velocity $\mathbf{v}_r$ , the terme $\mathbf{\Omega} \times \mathbf{u}$ is an apparent acceleration known as the Coriolis acceleration, which exists only if there is motion relative to a moving frame such as the earth. For this Coriolis acceleration we have

$$2\Omega \times \mathbf{u} = (2\Omega_O \cos\varphi\, w - 2\Omega_O \sin\varphi\, v)\mathbf{i} + 2\Omega_O \sin\varphi\, u\mathbf{j} - 2\Omega_O \cos\varphi\, u\mathbf{k}, \tag{2.13a}$$

when u, v, and w are the components of $\mathbf{u}$. The symbol $f = 2\Omega_O \sin\varphi$ is the local Coriolis parameter, and we can write, in place of (2.13a), the following relation:

$$2\Omega \times \mathbf{u} = f(u\mathbf{j} - v\mathbf{i}) + df/d\varphi(w\mathbf{i} - u\mathbf{k}). \tag{2.13b}$$

The importance of the Coriolis acceleration in relation to the inertial forces is given by the Rossby number Ro, which is defined as

$$\mathrm{Ro} = U_0/L_0 f_0, \text{with } f_0 = 2\Omega_O \sin\varphi_O, \text{ where } \varphi_O = \text{Constant}. \tag{2.13c}$$

In fact, $f_0 \approx 10^{-4}$ in mid-latitudes, and for $\mathrm{Ro} >> 1$, the Coriolis force is likely to cause only a slight modification of the pattern of atmospheric motion, but when $\mathrm{Ro} \ll 1$, the effects of the Coriolis force are likely to be dominant! In fact, for synoptic-scale atmospheric motions, we have: $L_0 \cong 10^6$ m and $\mathrm{Ro} \cong 10^{-1}$, but for the meso- (or regional) scale atmospheric flows, we have $L_0 \cong 10^5$ m and $\mathrm{Ro} \cong 1$. Finally, for the case of local-scale atmospheric process, we have rather $L_0 \cong 10^4$ m and $\mathrm{Ro} \cong 10$!

In the present book, we prefer to use, instead of the Rossby number Ro, the *parameter* (*taking into account my Moscovite (1957–1966) scientific meteo adventure with Il'ya Afanas'evich Kibel*) known as the *Kibel number*:

$$\mathrm{Ki} = (1/f_0)/t_0, \tag{2.14}$$

where $t_0$ is a characteristic reference time. If $t_0$ is the advective time scale $L_0/U_0$, this Kibel number is identical to the Rossby number Ro.

### 2.6.2 *The Atmosphere as a Continuum*

From a mathematical point of view of a continuum, we postulate that properties at any one point can be expressed in terms of properties at a neighboring point—this is because these properties and their derivatives are assumed continuous in their spatial variations. *From the continuum mechanics point of view, the atmosphere is a thin layer of air—a gaseous mixture—surrounding the surface of the earth, which remains attached to the earth by the pull of gravity and extends about seven miles upwards from the earth's surface.*

The layer nearest to the earth's surface, characterized by a linear temperature decrease with increasing altitude, is called the troposphere, and it is most influenced by energy transfer through radiation, evaporation, condensation, and convection. However, it is no easy matter to account for all these four effects. The troposphere also represents the limit within which conventional air flight takes place, and also within which man-made pollution from industrial activities is principally confined, and it is where most cloud formation occurs. Dynamically speaking, the troposphere is stable, but those portions of the layer nearest the surface of the earth are often unstable.

The air in the atmosphere is a Newtonian fluid, and dry air is governed by the law of perfect gases with (2.1) and (2.2d). If the relative velocities *are small*, the *pressure* will be only *slightly disturbed* from the value it would have in the absence of motion, $p_{st}(z_{st})$, defined by the relations

$$dp_{st}(z_{st})/dz_{st} + g\rho_{st} = 0, \text{with } \rho_{st}(z_{st}) = p_{st}(z_{st})/RT_{st}(z_{st}), \tag{2.15a}$$

in a basic, so-called *standard atmosphere* in which fluctuations due to motion occur. This basic standard atmospheric state is assumed known, although in fact its determination from first principles requires at least the consideration of mechanisms such as radiative transfer in the atmosphere. But in rather simple cases we have, from the first law of thermodynamics, for the standard temperature $T_{st}(z_{st})$, a *function only of the standard altitude* $z_{st}$, the following equation:

$$k(T_{st})\, dT_{st}(z_{st})/dz_{st} + R_{st}(T_{st}) = 0, \tag{2.15b}$$

with

$$dR_{st}(z_{st})/dz_{st} = \rho_{st} Q_{st}(T_{st}). \tag{2.15c}$$

Equation (2.15c) gives $R_{st}(T_{st})$, where $Q_{st}(T_{st})$ is the rate of heat supply per unit mass by radiative heat transfer and the scalar $k_{st}(T_{st})$ is the standard coefficient of *thermal conductivity*.

For our purposes here, it is sufficient to assume that $Q_{st}(T_{st})$ is a known function of $T_{st}$ and also that the influence of the rate of heat by radiative heat transfer on the atmospheric motions is the main factor in the determination of $T_{st}$. By doing this, we consider only a mean standard heat source and ignore variations thereof! As a consequence, the *reference quantities* are the values for the standard state at ground level: $p_{st}(0)$, $\rho_{st}(0)$, and $T_{st}(0)$. In this case, from (2.15a), the following non-dimensional parameter—the *Boussinesq* number—appears:

$$Bo = gH_0/(p_{st}(0)/\rho_{st}(0)) = H_0/(RT_{st}(0)/g), \tag{2.16a}$$

where $H_0$ is a characteristic length scale for vertical atmospheric motions, while the *characteristic altitude scale for the standard altitude* $z_{st}$ is

$$H_{st} = RT_{st}(0)/g \tag{2.16b}$$

In fact, *Bo is the ratio of two vertical length scales*. It follows from (2.16a) that, between the parameters Froude (Fr), Mach (M), Boussinesq (Bo), and the ratio ε—the so-called *hydrostatic parameter*—the following relation holds:

$$(Fr_{L0})^2 = \gamma\varepsilon M^2/Bo \ \text{ or } \gamma\,[M/Fr_{H0}]^2 = Bo, \tag{2.16c}$$

where

$$\varepsilon = H_0/L_0,\ \text{and}\ M = U_0/\sqrt{C_0},\ \ \text{with}\, C_0 = \gamma R T_{st}(0). \qquad (2.16d-f)$$

On the other hand, $N_{st}(z_{st})$ defined by

$$N_{st}{}^2(z_{st}) = [g/(T_{st})]\{[(\gamma-1)g/\gamma R] + dT_{st}/dz_{st}\}, \tag{2.16g}$$

is called the *Brunt-Väisälä frequency* or the *natural frequency of oscillations of a vertical column of "standard" atmospheric mass during a small displacement from its equilibrium position—the standard atmosphere being statically stable when* $N_{st}$ is *real*! Below, the existence of characteristic scales is exploited by the introduction of non-dimensional quantities denoted by primes, and we have

$$z = H_0 z',\ z_{st} = H_{st} z'_{st},\ \text{ and }\ T_{st} = T_{st}(0)T'_{st}. \tag{2.16h}$$

We can then write the following dimensionless equation in place of (2.16g):

$$a_{st}N'_{st}{}^2(z'_{st}) = [Bo/T'_{st}]\{[(\gamma-1)/\gamma] + dT'_{st}/dz'_{st}\}, \tag{2.17a}$$

where $a_{st}$ is a *dimensionless measure of the standard stability*. We can also derive an *interesting relation between* $z_{st}$ *and* $z'$, namely:

$$z_{st} = Bo\,z', \tag{2.17b}$$

which plays a decisive role in the justification (*à la Zeytounian (1974)*) of the famous *Boussinesq equations*. Concerning these *Boussinesq (shallow) equations*, and also *Zeytounian (deep) equations*, see Sects. 2.6.3 and 2.6.4 for their *asymptotic derivation from the Euler equations*, via the *RAM approach*.

Finally, in the asymptotics of atmospheric motions, two parameters also play an important role: the *Reynolds number* $\mathrm{Re} = \mathrm{U_0 L_0}/\nu_0$, which shows the *importance* of the *inertia relative* to the *viscosity*, and the *Ekman number*:

$$\mathrm{Ek} = \mathrm{Ro}/\mathrm{Re} = \left(1/\mathrm{L_0}^2\right)[\nu_0/f_0], \tag{2.17c}$$

which is a measure of the *ratio* of the *frictional* and *Coriolis* forces.

We observe that an important feature of *large synoptic-scale atmospheric motions* is that both the *Kibel and Ekman numbers are small.* A typical value of Ek in the earth's troposphere is $10^{-3}$, when for the eddy viscosity we choose $\nu_0 = 5\ \mathrm{m^2/s}$.

Except in the immediate vicinity of the equator, Ki is usually a small parameter ($\ll 1$), if the characteristic time scale $\mathrm{t_0} \gg 10^4$ s, which is the case for synoptic-scale motions.

*In any realistic atmospheric situation, M* $\ll 1$, *and the synoptic meteorological situation corresponds to*

$$\varepsilon \sim 10^{-2} \text{and Bo} << 1. \tag{2.17d}$$

But, frequently, for the *prediction of atmospheric phenomena at regional (meso) and local scales*, one may assume that

$$\mathrm{Bo} << 1, \text{ but } \varepsilon \sim 1, \tag{2.17e}$$

as is typically the case for *lee waves*, arising *downstream of a mountain.*

For further details concerning the physical nature of the atmosphere, the reader is referred to books *Houghton (1977)* and *Scorer (1978). It must be kept in mind that modelling, i.e., the translation of a complex physical situation into correctly expressed mathematical terms, has at the present time become very important in many fields in the realm of scientific research.*

### 2.6.3 Shallow Boussinesq Equations

To derive the *shallow Boussinesq equations from the Euler equations*, these Euler equations must be written in dimensionless form. First we take into account (2.15a) for the *standard atmosphere*, and with $p_{\mathrm{st}}(0)$, $\rho_{\mathrm{st}}(0)$, and $T_{\mathrm{st}}(0)$ we introduce non-dimensional quantities (dropping the primes) such that

$$p = p_{st}(z_{st})\,[1 + \pi], \quad \rho = \rho_{st}(z_{st})\,[1 + \omega], \quad T = T_{st}(z_{st})\,[1 + \theta]. \tag{2.18a}$$

With (2.18a), for dimensionless velocity components ($\mathbf{v} = (u, v)$, w), and thermodynamic perturbations, $\pi$, $\rho$, and $\theta$, we derive the following *dimensionless, emerging equations*:

$$\begin{aligned}
&[1 + \omega]\ \mathrm{St}\ D\mathbf{v}/Dt + T_{st}(z_{st})\left[1/\gamma M^2\right]\ \nabla\pi = 0,\\
&[1 + \omega]\ \mathrm{St}\ Dw/Dt + \left[1/\gamma M^2\right]\{T_{st}(z_{st})\partial\pi/\partial z\\
&\qquad\qquad - [1 + \omega]\ \mathrm{Bo}\ \theta\} = 0,\\
&\pi = \omega + [1 + \omega]\ \theta,\\
&\mathrm{St}\ D\omega/Dt + [1 + \omega]\ (\partial u/\partial x + \partial v/\partial y + \partial w/\partial z)\\
&\qquad = [1 + \omega]\ \mathrm{Bo}\ [1/T_{st}(z_{st})]\{1 - \Gamma_{st}(z_{st})\}w,\\
&\mathrm{St}\ D\theta/Dt - [(\gamma - 1)/\gamma]\ \mathrm{St}\ D\pi/Dt\\
&\qquad + [1 + \pi]\ \mathrm{Bo}\ [1/T_{st}(z_{st})]\{[(\gamma - 1)/\gamma] - \Gamma_{st}(z_{st})\}w = 0.
\end{aligned} \tag{2.18b}$$

Again we stress that the above dimensionless Euler meteo equations (2.18b) for $\mathbf{v} = (u, v)$, w, and ($\pi$, $\omega$, $\theta$), are a set of *exact consequences of the traditional Euler equations*, and this remark is very important for the consistent derivation of the Boussinesq model equations below!

Our approach via the RAM approach goes from the full exact equations to simplified/reduced model equations—and this is the only way to derive relevant and worthwhile reduced/simplified, non-ad hoc working model equations! As *leading-order* equations, these Boussinesq equations are derived via the RAM approach from the above Euler system of equations (2.18b), if we assume that

$$M << 1 \text{ and } \mathrm{Bo} << 1, \text{ such that } \mathrm{Bo}/M = B^* = O(1), \tag{2.18c}$$

and use the associated asymptotic expansions

$$\begin{aligned}
(u, v, w) &= (u_{0B}, v_{0B}, w_{0B}) + \ldots,\\
(\omega, q) &= M\ (\omega_{1B}, \theta_{1B}) + \ldots,\\
\pi &= M^2\ \pi_{2B} + ,
\end{aligned} \tag{2.18d}$$

and also the limiting process:

$$\lim_{M\to o}[\text{with t, x, y, z, and St, } \gamma, \ B^* \text{ fixed}]. \tag{2.18e}$$

The above conditions (2.18c)–(2.18e) are obtained after a brief investigation of the various degeneracies of the Euler equations (2.18b) when $M \to 0$ with t, x, y, z, and St, $\gamma$, and $B^*$ fixed! In this case, it appears that one must assume that the *Boussinesq number Bo is also a small parameter* such that the *similarity relation with $B^* = O(1)$*, in (2.18c), is satisfied!

From the above conditions (2.18c)–(2.18e), we derive the following Boussinesq model equations for the functions: $u_{oB}$, $v_{oB}$, $w_{oB}$, $\omega_{1B}$, $\theta_{1B}$, and $\pi_{2B}$ in place of the Euler equations (2.18b) (written without primes):

$$\begin{aligned}
&\mathrm{St}\, \mathrm{D}\mathbf{v}_{0B}/\mathrm{Dt} + [1/\gamma]\ \nabla\pi_{2B} = 0,\\
&\mathrm{St}\, \mathrm{D}w_{0B}/\mathrm{Dt} + [1/\gamma]\ \partial\pi_{2B}/\partial z - B^*\theta_{1B} = 0,\\
&\omega_{1B} = -\theta_{1B},\\
&\partial u_{0B}/\partial x + \partial v_{0B}/\partial y + \partial w_{0B}/\partial z = 0,\\
&\mathrm{St}\, \mathrm{D}\theta_{1B}/\mathrm{Dt} + B^*\{[(\gamma - 1)/\gamma] - \Gamma_{st}(0)\}w_{0B} = 0,
\end{aligned} \tag{2.18f}$$

since in dimensionless form, we obviously have $T_{st}(o) \equiv 1$, but in a *general case* $\Gamma_{st}(0) = dT_{st}(z_{st})/dz_{st|z_{st}} = 0$ is obviously *different from zero!*

We note also that the above relation (2.17b) was taken into account. Concerning the validity of the above system of *Boussinesq equations,* from the similarity relation (2.18c), where $B^* = O(1)$, we have

$$B^* = \mathrm{Bo}/\mathrm{M} = \mathrm{O}(1) \Rightarrow \mathrm{H_c} \approx [\mathrm{U_c}/g](\mathrm{RT}_{st}(0)/\gamma)^{1/2} \equiv H_B \approx 10^3 m, \tag{2.18g}$$

and this leads to a strong restriction on the application of the Boussinesq equations (2.18f) in the whole troposphere ($\mathrm{H_c} = \mathrm{H}_{st} = \mathrm{RT}_{st}(0)/g \approx 10^4$ m)!

The equations (2.18f) derived above are indeed only *shallow Boussinesq equations*. Concerning more particularly the *lee waves downstream of a mountain in the whole troposphere*, one must assume that $\mathrm{Bo} \equiv 1$, $z_{st} \equiv z$. In the 2003 paper <8>, the reader can find a more general derivation of the Boussinesq equations applicable to atmospheric motions

### 2.6.4 Deep Equations "à la Zeytounian"

If we take into account that, *in the troposphere, the air temperature on average decreases with height at an overall positive rate (of about 6.5 °C/km)*, then it seems to me not too bad a hypothesis to assume that $-\Gamma_{st}(z) = -dT_{st}/dz$ *is very close to* $[(\gamma - 1)/\gamma]$ *in the troposphere*, and write

$$\Gamma_{st}(z) = -\, dT_{st}/dz = [(\gamma - 1)/\gamma] + M^2\chi_{st}(z), \tag{2.19a}$$

the function $\chi_{st}$ (z) being a known function of z which takes into account a *weak stratification with altitude* z in a *standard troposphere*, and $|\chi_{st}(z)| = O(1)$.

In such a case, with

$$\lim_{M\to 0}[\text{with t, x, y, z, and St, } \gamma, \text{ Bo, } \chi_{st}(z) \text{ fixed}], \tag{2.19b}$$

*instead of the above system* (2.18f) *of shallow Boussinesq equations, we derive* the following so-called *deep equation* (*à la Zeytounian*) (see Sect.10.2.2 in [22], cited in "Introduction"):

$$\begin{aligned}
&\gamma\ T_d(z)\ [\partial u_{od}/\partial x + \partial v_{od}/\partial y + \partial w_{od}] = w_{od},\\
&\mathrm{St}\ Dv_{od}/Dt + (1/\gamma)\ T_d(z)\ \nabla\pi_{2d} = 0,\\
&\mathrm{St}\ Dw_{od}/Dt + (1/\gamma)[T_d(z)\partial\pi_{2d}/\partial z - \theta_{2d}] = 0,\\
&\mathrm{St}\ D\theta_{2d}/Dt - [(\gamma - 1)/\gamma]\ \mathrm{St}\ D\pi_{2d}/Dt + [1/T_d(z)]\chi_{st}(z)w_{od} = 0,\\
&\pi_{2d} = \theta_{2d} + \omega_{2d},\ T_d(z) = 1 - [(\gamma - 1)/\gamma]\ z,
\end{aligned} \tag{2.19c}$$

with $D/Dt = \partial/\partial t + u_{od}\partial/\partial x + v_{od}\partial/\partial y + w_{od}\partial/\partial z$, when $M \to 0$ with t, x, y, z, and St, $\gamma$, Bo $\equiv 1$, fixed, with:

$$\begin{aligned}
&(u,\ v,\ w) = (u_{0d},\ v_{0d},\ w_{0d}) + \ldots,\\
&(\omega,\ \theta,\ \pi) = M^2(\omega_{2d},\ \theta_{2d},\ \pi_{2d}) + \ldots.
\end{aligned} \tag{2.19d}$$

The above model equations (2.19c) are the true "*anelastic*" non-dissipative equations, the equations of *Ogura* and *Phililips (1962), being rather an ad hoc system of equations!*

These *deep convection equations* (2.19c) *are valid in the whole troposphere, while the shallow Boussinesq equations* (2.18f) *can be considered as a reduced form of these deep convection equations* (2.19c), *only valid in the vicinity of the ground, in a layer close to a flat ground surface with a thickness of only about* $10^3$ m.

In the *2D steady case*, from *each of the above systems of equations*—(2.18f) and (2.19c)—one can derive a *single equation for a stream function*—a very pleasing derivation for a curious and motivated student!

After Chap. 6, in "*Some Concluding Remarks about Part III*", the reader can find the *RAM derivation* of a more complete *anelastic, deep, non-adiabatic, viscous and heat conducting system of equations for dissipative atmospheric thermal convection.*

## 2.7 Complementary Remarks

In our books published by *Springe*r (Heidelberg) during the years 1974–2014, the reader can find many complementary presentations of very different theoretical subjects relating to fluid dynamics—the reader will find some of these below! But first it should be noted that much of the impetus for research on *Newtonian fluid dynamics* during the past 50 *years* was created by the rapid development of "*asymptotics*" and "*modelling*".

In [10], cited in "Introduction" and in the paper by *Guiraud* [7], cited in Chap. 1 entitled *Going on with asymptotics*, and also in my 2002 book, *FMIA 64* [7], cited in "Introduction" (pp. 4–18 contain a *short summary of* Chaps. 2–12), the reader can find discussion of the following: *Newtonian fluid flow equations and conditions*; *asymptotic analysis and modelling*; *useful limiting forms of the NSF equations*; *the Navier-Fourier model*; *the Euler mod*el; *boundary layer models*; *models of nonlinear acoustics*; *low-Reynolds number asymptotics; asymptotic modelling of thermal convection and interfacial phenomena; meteo fluid dynamics models; singular coupling and the triple-deck model.*

On the other hand, in the book by *R. E. Meyer* <*9*>, the curious reader will find an *introduction to mathematical fluid dynamics* laying out the *basic concepts* of a *semi-axiomatic foundation,* but *without abstract nonlinear functional analysis,* in contrast to the two books by *P. L. Lions, published in 1996 and 1998* by Oxford University Press, *where pure mathematics plays the main role.* We note that, *for a mathematics student, such a treatment (à la Meyer) helps to dispel the common impression that the whole subject is built on a quicksand of assorted intuitions. It remains to hope that our RAM approach will lead to more positive feelings!*

*Of course, in Meyer's book, the account is not axiomatic; the "postulates" are used to illuminate the subject, not to deduce it, and* mathematicians will be even more disappointed by the lack of any attempt to prove an existence theorem, or even to talk seriously about partial differential equations. There would seem to be room, however, for an introduction to what the whole subject is about, before the Navier-Stokes equations are tackled. In fact, many mathematicians will complain that the book contains hardly (!) any mathematics—which is also the case in the present book—and many theoretical fluid dynamicists, that the book is far too abstract—*this indicates a gap which should be filled*!

Note, however, that, in Sect. 38 of Chap. 6 of Meyer, 1971, the reader can find a *short but well argued discussion* concerning the Navier-Stokes equations, referred to in the present book as the *Navier-Stokes-Fourier (NSF) equations.* In particular, in pp. 100–110 of Chap. 4 of Meyer, 1971, the reader will find *the formulation, solution, and discussion of the basic classical Blasius problem of the theory of fluids with small viscosity, considering a steady incompressible flow past a solid flat plate placed edgewise in a uniform stream.* The Blasius problem for viscous fluid flow is considered for the case of a slightly compressible fluid in Sect. 8.5.

Chapter 5 of Meyer, 1971, entitled *Some aspects of rotating fluids*, gives useful complementary infor*mation about* Sect. 2.6 *of the present book, and the singular perturbation example* in his Chap. 4 *seems to me a good simple case, illustrating asymptotics!*

In our two volumes [13] and [14], cited in "Introduction", the reader will find many and varied examples of the application of *asymptotics* to both *non-viscous* and *viscous* fluid dynamics problems. In particular, *in* Chap. 3 of [14], cited in "Introduction", the reader will find some *simple* (but *fundamental*) examples of viscous fluid flows.

For instance, Sect. 3.1 considers the *plane Poiseuille* flow and the *Orr-Sommerfeld* equation, which is solved by a *double-scale asymptotic technique* when

$$ik\text{Re} = 1/\varepsilon >> 1. \tag{2.20a}$$

The unknown stream function ψ(t, x, z), satisfying the perturbed vorticity Navier equation, is written in the following form:

$$\psi(z) = \phi(z) \exp[ik(x - ct))], \tag{2.20b}$$

and ϕ(z) is the solution of the *Orr-Sommerfeld* equation (see p. 58–60 of [14], cited in "Introduction"). With the boundary conditions $d\phi/dz = 0$ and $\phi = 0$ on $z = -1$ and $z = +1$, this is an *eigenvalue problem.*

Writing $c = c_r + ic_i$, then if $c_i = 0$, there is sustained oscillation, and the condition $c_i = 0$ leads to a relation between k and Re—a *neutral stability curve* in the (*k*, Re) plane, as shown in Fig. 2.1.

Section 3.2 of [14], cited in "Introduction", is devoted to *steady flow* through an *arbitrary cylinder under pressure,* and three particular cases are considered—the case of a *circular* cylinder, the case of *an annular region* beween *concentric cylinders*, and the case of a cylinder of *arbitrary section.*

Section 3.3 of [14], cited in "Introduction", *investigates* the problem of the *steady-state Couette flow between two concentric circular cylinders of radii* $r_1$, $r_2$, rotating about their common axis at angular velocities $\omega_1$, $\omega_2$. This leads to the *classic Taylor problem*, where the *Taylor number for a small* $\varepsilon = d/r_m \ll 1$ is

$$\text{Ta} = (\rho^\circ/\mu^\circ)^2 (\omega_1)^2 d^4 \left[(1 - \Omega^2)/\varepsilon\right], \tag{2.21a}$$

and the ratio $(1 - \Omega^2)/\varepsilon$, of two small parameters is assumed to have a finite value. In $\varepsilon = d/r_m$, $d = r_2 - r_1$, and $r_m = (½) [r_2 + r_1]$, while $\Omega = \omega_2/\omega_1$. Again we have an eigenvalue problem that yields an eigenvalue relation, and in Fig. 2.2 the neutral

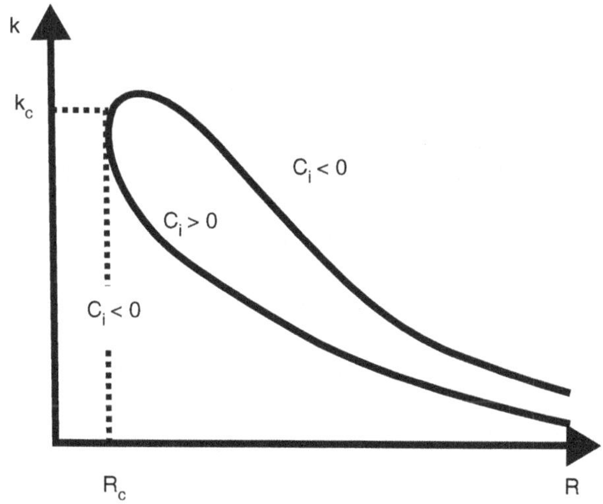


**Fig. 2.1** Neutral curve for plane Poiseuille flow. From p. 59 of [14], cited in "Introduction"

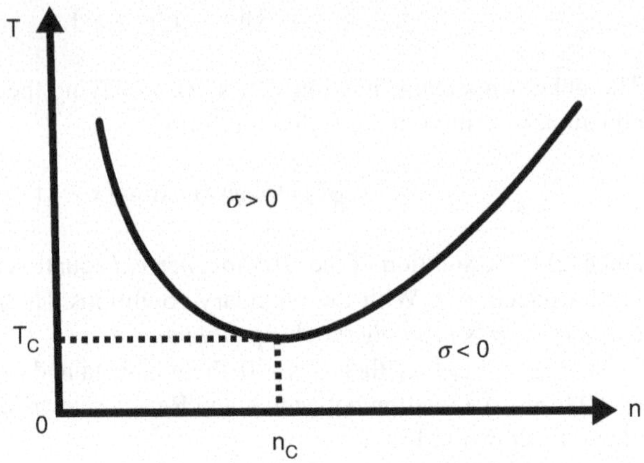


**Fig. 2.2** Neutral curve for circular Couette flow. From p. 67 of [14], cited in "Introduction"

stability curve is given by the value $\sigma = 0$. This is the *principle of exchange of stabilities*!

The flow is described by

$$u = w = 0,\ \ v = V(r) = Ar + B/r,\ \ (1/\rho^\circ)dP(r)/dr = (1/r)\,[V(r)]^2, \quad (2.21b)$$

and, following *Taylor (1923)*, it is assumed that the perturbed flow has the form $\mathbf{u} = (u', V(r) + v', w'), p = P(r) + p'$. In Fig. 2.2, n = *md with m the axial wave number.*

The Bénard problem is discussed in pp. 68–70 of Sect. 3.4 of [14], cited in "Introduction", as in Sect. 6.4 *of the present, but in the framework of a linear theory, when perturbations relative to a basic equilibrium state are small. In such a case, using the usual linearization, we derive a linear system à la Boussinesq. When we analyze the perturbations into normal mode,* we must then consider, for the *function W(z),*

$$(w',\ T') = [W(z),\ \Theta(z)]\ f(x,\ y)\exp(\sigma\ t), \quad (2.22)$$

where $\sigma = \sigma_r + i\sigma_i$. *In this case, the equation* for *W(z)* with the *boundary conditions determines a so-called self-adjoint eigenvalue problem for Rayleigh number,* $Ra = Pr\,Gr$, *where according to (3.7)* $Gr = \alpha/Fr_d^2$ *is the Grashof number with α as a small dilatation parameter* (see the Sect. 6.4).

When Ra exceeds the critical value $Ra_c$, instability occurs in the form of convection in (*Bénard*) *cells* forming a polygonal platform. Once again, the case $\sigma = 0$ represents neutral (marginal) stability (see Fig. 2.3).

The *formation of Bénard cells* in a weakly expansible liquid layer is one of the most remarkable examples of a *bifurcation phenomenon. Bifurcations in dissipative (dynamical) systems are investigated in* Chap. 10 (entitled *A Finite-Dimensional*

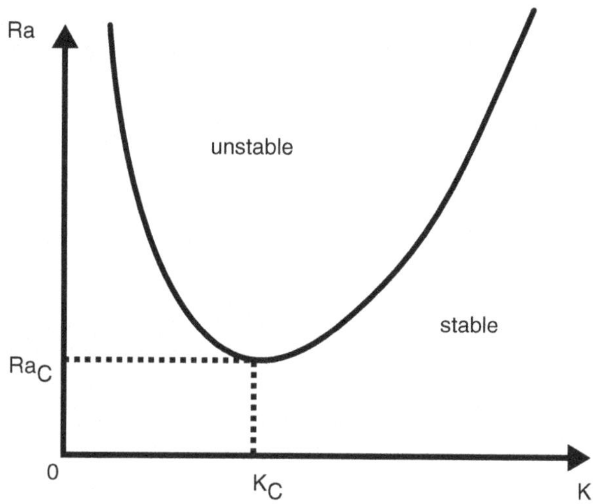


**Fig. 2.3** The neutral stability curve for the onset of Bénard thermal convection. From p. 70 of [14], cited in "Introduction"

*Dynamical System Approach to Turbulence) of the book* [14], cited in "Introduction", devoted to *the theory and applications of viscous fluid flows.*

The linear problem with a *free surface* open to *ambiant passive air* is considered in Sect. 3.5 *of* [14], cited in "Introduction". This problem represents one of the most important cases in which capillary forces come into play. More precisely, the motion induced by *tangential gradients* of variable (only *temperature dependent*) *surface tension:*

$$\sigma(T) = \sigma(T^{\circ}) - \gamma^{\circ}\,(T - T^{\circ}), \tag{2.23a}$$

where $T^{\circ}$ is the constant temperature of the free surface in the basic equilibrium state

$$\gamma^{\circ} = -\,(\mathrm{d}\sigma/\mathrm{dT})_{T^{\circ}} = \text{const}, \tag{2.23b}$$

is customarily called the *Marangoni* effect.

But, according to Zeytounian (see Chap. 7 of [25] cited in "Introduction"), from an *asymptotic* viewpoint, *it is not consistent to take into account the buoyancy effect and the deformation of the free surface simultaneously* in the Bénard thermal convection model problem for a weakly expansible liquid!

In the *Bénard-Marangoni thermocapillary problem, three non-dimensional parameters play a significant role:*

$$\mathrm{Ma} = \gamma^{\circ}\mathrm{d}^{\circ}\left[\Delta T^{\circ}/\rho^{\circ}\nu^{\circ 2}\right], \tag{2.24a}$$

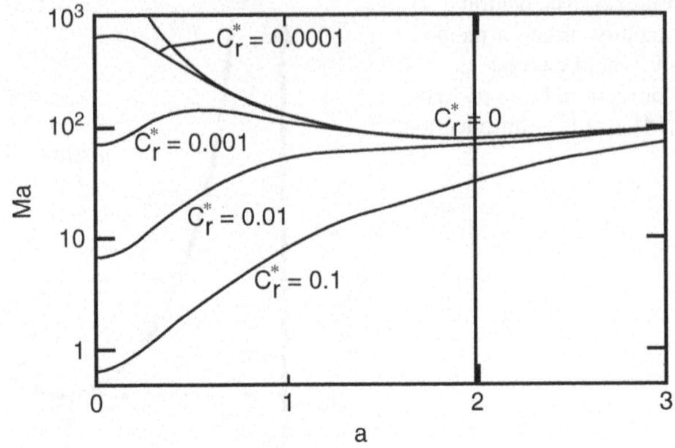


**Fig. 2.4** Neutral stability curves for various values of Cr (for Bi = 0) and Bn = 0.1. From p. 74 of [14], cited in "Introduction"

$$\mathrm{We} = \sigma(T^\circ)\left[\mathrm{d}^\circ/\rho^\circ \nu^{\circ 2}\right], \tag{2.24b}$$

$$\mathrm{Bi} = \mathrm{q}^\circ \mathrm{d}^\circ / k^\circ, \tag{2.24c}$$

which are respectively the *Marangoni, Weber, and Biot numbers. Here* $\Delta T^\circ$ is a difference of constant temperatures, and $q^\circ$ in the expression for the Biot number is the *thermal surface conductance constant* between a free surface and air.

Figure 2.4 shows the neutral stability curves for various values of the *crispation number*, according to the paper by *Regnier and Lebon (1995)*. For Bi = 0 and *Bond number* $\mathrm{Bn} = \mathrm{Pr}\,[\mathrm{Cr}/\mathrm{Fr}^{\circ 2}] = g\rho^\circ[\mathrm{d}^{\circ 2}/\sigma\,(T^\circ)] = 0.1$, the regions below each of curves represent the stable state in the plan (a, Ma). When $\mathrm{Cr} = \mathrm{We}/\mathrm{Pr} \to 0$, i.e., for a *non-deformable free surface*, and *Froude number* $\mathrm{Fr}^{\circ 2} \to 0$, such that $\mathrm{Bn} \neq 0$, we obtain once again the result due to *Pearson (1958)*, viz., $\mathrm{Ma_c} \approx 80$ and $\mathrm{a_c} \approx 2.0$!

Concerning Sects. 6 and 7 in Chap. 3 of [14], cited in "Introduction", we note the following. *First*, in Sect. 6, viscous flow ($\nu^\circ$ is the kinematic viscosity coefficient) due to a disc (in the plane z = 0) rotating at constant angular velocity $\Omega^\circ$ is considered for small and large values of $\zeta = [\nu^\circ/\Omega^\circ]^{1/2}$ z. Then, in Sect. 7, the corresponding expansions are "joined" in a rather heuristic manner using a trial and error approach.

*Second*, in Sect. 7, the *Rayleigh flow problem* is investigated using the one-dimensional unsteady-state NSF equations, caused by an impulsively started flate plate, when

$$\mathrm{Pr} = 1,\ \mathrm{Re} = \mathrm{O}(1),\ \mathrm{M}^2 \to 0,\ \text{and}\ \tau^\circ \to 0, \tag{2.25a}$$

where $\tau^\circ = \Delta T^\circ/T^\circ$, and we assume that, for t > 0 on z = 0, the temperature changes *instantaneously* from $T^\circ$ to $T^\circ + \Delta T^\circ$, with $\Delta T^\circ > 0$. We first consider the *small* $M^2$ *asymptotic* solution *close* to the flat plane, but *far from* the initial time, which is an *inner* expansion. Then we examine the *small* $M^2$ *asymptotic* solution far from a flat plate, an *outer* expansion. As a result, it appears necessary to include a term proportional to $M^3$ in the expansions as a consequence of matching. However, the above procedure yields only an asymptotic solution for the time $t = O(1)$, and is not valid near $t = 0$, because the above asymptotic solutions would imply that an infinite impulse per unit area is required to set the flat plate in motion!

The main problem is to match the solution valid in the initial transient layer with the above asymptotic solution, which is valid for fixed time $t > 0$. In fact, for *small* $M^2$ *close to* the initial time, one must first introduce a *short time* adapted to the *initial transient layer. Hence, we write*

$$t^* = t/M^{\beta}, \ \beta > o, \tag{2.25b}$$

and with this short time (2.25b), we consider an *initial limiting process*:

$$M \to 0 \text{ with } t^* \text{ and z fixed and } Re = 0(1), \tag{2.25c}$$

with an *initial asymptotic expansion. If we assume that, close* $t^* = 0$, the *significant* system of equations is the *classical system of acoustic equations, then in the initial asymptotic expansion (valid close to* $t^* = 0$*) we have a term* proportional to the Mach number M, and $\beta = 1$ in (2.25b).

In *Some Concluding Remarks for Part I* below, the reader can find further information concerning this unsteady *Rayleigh flow problem.*

As a conclusion of this rather long Sect. 2.7, I wish to quote below some comments by *M.F. Platzer* (from the Dept of Aeronautics and Astronautics at the Naval Postgraduate School, Code *AA/PL*, Monterey CA 93943-5000) in the Book Reviews section of *Appl Mech Rev*, vol 57, No. 3, May 2004. Platzer reviewed my book entitled *Theory* and *Applications of Viscous Fluid Flows* [14], cited in "Introduction":

> It is evident from this brief summary that the author's emphasis is on the mathematical aspects of the viscous flow equations and their various asymptotic limit cases and analytical solution methods.
>
> His choice of topic and flow problems is meant to provide young researchers in fluid mechanics, applied mathematics and theoretical physics with an up-to-date presentation of some key problems in the analysis of viscous fluid flows.
>
> Although the author intentionally limited himself to a select few topics, teachers of advanced viscous flow courses and researchers in this field will welcome this book for its thorough review of current work and the listing of 1156 relevant papers.
>
> In my judgment, it meets the stated objective of bridging the gap between standard undergraduate texts in fluid mechanics and specialized monographs.

The last chapter (Chap. 10) of our *Theory and Applications of Viscous Fluid Flows* [14]cited in "Introduction" presents the *finite-dimensional dynamical systems approach to turbulence* by reviewing the classical *Landau-Hopf, Ruelle-Takens-Newhouse, Feigenbaum,* and *Pomeau-Maneville transition scenarios to turbulence*. In pp. 414–443 of Sect. 10.4 of [14], cited in "Introduction" the reader will find a collection of *strange attractors (see* Figs. 2.5 and 2.6) for various viscous flow phenomena. These two strange attractors are related to the thermocapillary instabilities in a free-falling vertical 2D film, when the amplitude KS equation (2.26) is considered, where the function H(t, x) is related with the film thickness.

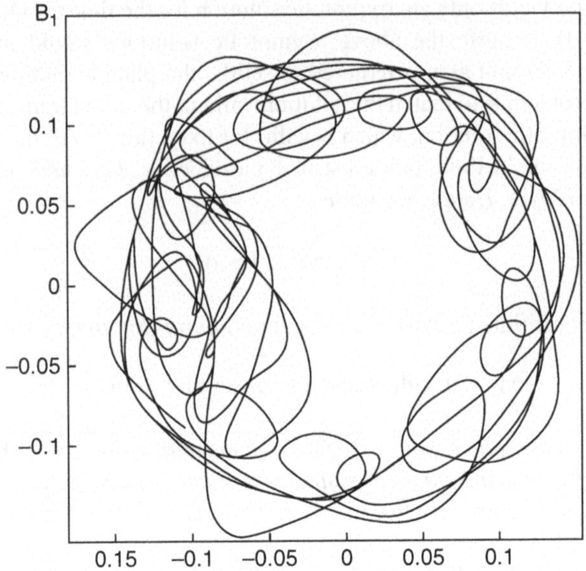


**Fig. 2.5** Metastable chaos regime at k = 0. 273. From p. 424 of [14], cited in "Introduction"

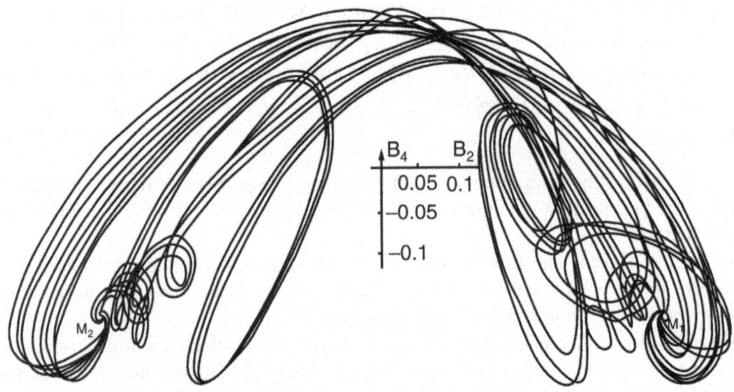


**Fig. 2.6** Complex movement is observed in the vicinity of limit cycles. From p. 423 of [14], cited in "Introduction"

$$\partial H/\partial t + 4H\partial H/\partial x + \partial^2 H/\partial x^2 + \partial^4 H/\partial x^4 = 0. \tag{2.26}$$

# References


<1> Rayleigh Lord (1916) On convection currents in horizontal layer of fluid when the higher temperature is on the under side. Philos Mag Ser 6 32:529–546 (192)

<2> Coulouvrat F (1992) On the equations of nonlinear acoustics. J Acoustique 321–359, 5

<3> Zeytounian RKh (1991) Meteorological fluid dynamics. LNP m5. Springer, Heidelberg

<4> Zeytounian RKh (1968) Études Hydrodynamiques des Phénomènes Méso -météorologiques. Cours à l' École de la Météorologie(pp. 163). Direction de la Météorologie—Quai Branly à Paris

<5> Zeytounian RKh (1985) Recent advances in asymptotic modelling of tangent atmospheric motions. Int J Eng Sci 23:1239–1288

<6> Zeytounian RKh (1991) Fluid dynamics limits of meterological equations. Adv Mech 14(3):3–86 (Warszawa)

<7> Zeytounian RKh (2004) Weather prediction as a problem in fluid dynamics (On the 100th birthday of I A Kibel). Izv Atmos Ocean Phys 40(5):593–605

<8> Zeytounian RKh (2004) On the foundations of the Boussinesq approximation applicable to atmospheric motions. Izv Atmos Ocean Phys 39:S1–S14

<9> Meyer RE (1971) Introduction to mathematical fluid dynamics. Wiley, New York, NY. Re-edition by Dover Publications

# Chapter 3
# A Basic Postulate and Some Key Steps of the RAM Approach

In the RAM approach, all terms in the NSF system of equations and conditions are significant and, as a consequence, it is strictly forbidden to carry out any a priori simplification. In the RAM approach we use the following terms:

Rational—This means that a logical method is used in order to obtain from the full NSF initial-boundary value problem a significant, non-contradictory, logically consistent approximate working model subject to numerical simulation.

Asymptotics—This insists on the main role devoted to matched asymptotic expansions (MAE) produced by various limiting processes. It at least gives the possibility of obtaining various approximate leading-order models—but these can be improved by asymptotic expansions leading to associated second-order models which take into account—from the full NSF problem—some of the terms neglected in the leading-order models.

Modelling—This is probably the most important of the three terms. It stresses the fact that the ultimate goal is to construct approximate working models that are more easily accessible to numerical simulation/computing than the original one consisting of the full unsteady NSF system of equations with initial and boundary conditions.

Here the word rigorous does not have the significance attached to it in pure mathematics, and it is not strictly speaking the failure to obtain existence and uniqueness that makes a significant difference between our RAM approach and abstract—or so-called "*pure*"—mathematics.

The RAM approach leads naturally to numerical simulation and, through it, to results of obvious practical interest, which is almost equivalent to stating existence and uniqueness. We may emphasize here the various technologically and geophysically useful applications of such numerical results. In what follows I intend to explain in a completely logical way the process underlying the RAM approach—but of course, "logical" does not mean that we give a proof of existence and uniqueness.



R.Kh. Zeytounian, *Challenges in Fluid Dynamics*,
DOI 10.1007/978-3-319-31619-2_3

*In fact the main goal is to decompose the global space-time domain where the original NSF initial-boundary value (NSF I-B V) problem has been formulated into several localized sub-domains.* Within each of these, a reduced, approximate, but simplified working model problem has to be solved, generally by numerical simulation/computation. Then the whole set of these solutions must be *put back together* in order to give an approximate global solution to the original (NSF I-B V) problem.

Limiting processes and matching both rely on asymptotics. The counterpart is that each model is only approximate and the recomposed numerical result seems to be even more appropximate than each of its parts, due to matching. Nevertheless it may be possible to continue each expansion and eventually to take account of certain complementary physical effects ignored in the leading-order global result.

We should also mention here that matching cannot always be achieved, due to some incompatibility, and this precise point will be considered in Sect. 8.9, devoted to application of a more subtle process than matching, namely *compatibility-non-secularity conditions.*

Let us go beyond sterile polemic towards a reconciliation of the two approaches: logical and significant non-contradictory *rational asymptotic modelling*, on the one hand, and estimation of error through the use of certain norms supplied by *abstract nonlinear functional analysis*, on the other hand. The aim is to escape from the rather obvious but ultimately void dispute. Indeed, fluid dynamics is an interesting and many-faceted discipline which today poses various unresolved fundamental, practical, theoretical, numerical, and mathematical problems.

In particular, mathematicians should be *firmly persuaded* that, thanks to the RAM approach, they *have here a way to utilize valuable working models as starting problems* for their rigorous mathematical investigations, these being derived as *significant degeneracies* of the NSF system of equations! Concerning numericians in particular, it seems judicious to mention a sound judgement by the late D. Euvrard from his abstract (pp. 25–26 of the 1986 paper in [10]) entitled *A mutually profitable alliance: asymptotic expansions and numerical computations*. Euvrard wrote: "We intend to convince the reader that both schools should not compete or even ignore each other, but rather take advantage of working together. We even think that any team of numericians should devote part of its efforts to asymptotic or analytical work." And at the end of his paper, before the list of references, he added: "*Does the reader still fear that modern numerical computations will kill traditional analytical research?*"

Personally, *I fear* rather that so-called mathematical fluid dynamics *will kill traditional fluid dynamics modelling* which, via the RAM approach, seems to present a *good case. Indeed, the latter approach deserves more serious computational attention than it has received*!

Numerical simulations/computations of complicated real fluid flows is an expanding activity, and costs a lot of money! What we suggest here is simply the following: *before entering directly into numerics for a given simulation/computation NSF fluid flow problem, one should ask whether it is not possible to reduce the original/starting problem into a few others, intimately connected, and easily reassembled in order to provide, after (ultimately necessary) numerical simulation, a global approximate solution! Naturally, this will appeal to numericians. Why shouldn't theoretical fluid dynamicists and corresponding numericians become acquainted with one another?*

With Figs. 5.1 and 5.2 in Chap. 5 and Figs. 6.1, 6.2, 6.3 and 6.4 in Chap. 6, the reader will find a variety of valuable models, all derived in a logical way from the full NSF system of equations. The basic argument rests on a hypothesis which will be expressed quite simply in Sect. 3.1, namely, any derivative occurring in the NSF equations has, for any practical problem, a numerical value which varies over a limited range of the real numbers, within some sub-domain of the original one over which the NSF system of equations is valid. Such a hypothesis is converted into the possibility of applying calculus with the help of dimensionless analysis, and this leads to a number of non-dimensional reference parameters. A list of the main such parameters is given at the beginning of Sect. 3.2.

Having done this, the whole of the work rests on our RAM approach, formulated in our companion 2012 book. Here, in Sects. 3.1–3.4, the reader will find a discussion of the way this RAM approach is actually carried out!

## 3.1 The Basic Postulate of the RAM Approach

To present the RAM approach, we start from the beginning *with a basic assumption* that we consider as a *postulate*:

> If to leading order, an approximate model is derived from an I-BV NSF fluid flow problem, then a RAM approach must be adopted to make sure that terms neglected in such a full unsteady NSF stiff problem, really are much smaller than those retained in the derived approximate, simple—but consistent—non-stiff, leading-order, working model problem.

*This "basic postulate", a statement that is accepted as true, is in fact a good general rule, despite its simplicity and maybe its naivety.* It is a statement of what the *fluid dynamicist* is advised to do in various modelling situations. For the *realization* of the *postulate* formulated above using our *RAM approach*, we must first implement several key steps listed in Sect. 3.3. These play a decisive role in the rational asymptotic derivation of consistent working models, as is clearly indicated in Sect. 3.4, in the framework of *the application of the RAM approach to atmospheric motions.*

In fact, *asymptotics* together with the *matching* and *limiting processes* are the two main tools involved in "*deconstruction*" via the RAM approach! In Sect. 5.1 of Chap. 5, which appears in Part II, we give some information concerning this "*deconstruction*" *process.* It is achieved mainly via the ploy of considering the various NSF problems in dimensionless forms, the consequence of such an approach being the appearance of a family of *non-dimensional parameters* (*numbers*)!

## 3.2 Typical Dimensionless NSF Problem and Reduced Parameters

*Low viscosity fluids constitute a crucial part of the application of asymptotics to fluid dynamics problems. In* Sect. 6.1, *the reader will find a RAM-deconstruction approach in the framework of the full unsteady NSF (I-B V) problem.*

In fact, the practical usefulness of the *inviscid (non-viscous) fluid concept* derives from the fact that the most commonly encountered fluids have *low viscosity* and *slip over a solid surface*. More precisely, if we consider the *normal velocity* of a regular surface *V* which is a piecewise continuous function of position on a regular surface—then we write:

$$(\mathbf{u} - V)\ .\ \vec{\mathbf{n}} = 0, \tag{3.1}$$

where $\vec{\mathbf{n}}$ is the unit normal to regular surface.

That is, *the normal component of the relative fluid velocity is zero at very point of a regular surface moving with the fluid.* Although it has long been known to some physicists (according to Sect. 18 in the book <9>, cited in Chap. 2 by *Meyer mentioned above*, devoted to *kinetic theory*), the fluid dynamical literature has but recently acknowledged that the *kinetic explanation* of viscous shear must have immediate *implications* for the *behavior of a gas near a solid surface. Hence*, according to Meyer, we must adopt as final conclusion: "*The boundary condition at a solid surface is zero relative fluid velocity, both with respect to the tangential and normal velocity components.*"

Of course, *solid boundaries are not the only possible boundaries of a fluid flow*! But a discussion of all boundary conditions relevant to Newtonian fluids would lead too far afield, and we shall be content (like Meyer <9>, cited in Chap. 2!) with the above rather strange result. In fact, the kinetic model implies inescapably that the inviscid fluid can have strictly *no slip* at a solid surface—in fact, one must take into account the fact that, close to a solid boundary, we also have a *Knudsen layer* of (relative) thickness M/Re, while $\mathrm{Re}^{-1/2}$ is the thickness of the Prandtl boundary layer.

This conclusion will obviously appear quite *absurd* (!) to most readers, since it contradicts the well-founded statement that an *inviscid fluid is free to slip at a solid surface*! It is also *appears inconsistent* with the *mathematical requirement of well-posedness!* This "*absurd*" conclusion was known to *Prandtl as early as 1904*, and would imply *rejection of the basic kinetic model (?)*. But it is in fact just another form of the *ideal fluid paradox*!

*As a first basic key step* in the RAM approach, we work with dimensionless quantities, and such a strategy is very important because it allows us to have at hand various *non-dimensional reference parameters*, which are *physical constants*. These are very good indicators of the *various physical effects* in the unsteady NSF equations: *unsteadiness, viscosity, compressibility, gravity, and so on*, and also in the boundary conditions, e.g., the *wall temperature* effect! Hence, $p$, $\rho$, and

$T$ are reduced relative to their *constant values* $p_c$, $\rho_c$, and $T_c$, where $p_c = R\rho_c T_c$ in the case of a thermally perfect gas. Then the components (u, v, w) of the velocity vector **u** are non-dimensionalized with constant velocity $U_c$. Regarding the length scales for the Cartesian coordinates (x, y, z) of the position space vector **x**, and time t, we choose, respectively, $L_c$ and $t_c$; coefficients, $\lambda$, $\mu$, and $k$ are assumed constant in equations (2.2a) and (2.2c).

However, it is important to observe that the above considerations concerning the dimensionless quantities are *not usually valid throughout the whole of the relevant fluid flow domain.* In fact, in the various *localized regions*, the *NSF-I-BV dimensionless problem is* of*ten singular* in the relevant fluid flow domain. *This is typically the case near the initial time t = 0 (as a consequence of filtering acoustic waves) and in the vicinity of the wall of a given body in movement* (for a *vanishing viscosity with a no-slip condition on the wall of the body*).

In the case of equations (2.2a–2.2c), with (2.2d and 2.2e), and conditions (2.10a–2.10d), the corresponding dimensionless quantities are introduced:

$$t^* = t/t_c, \quad (x^*, y^*, z^*) = (x, y, z)/L_c, \tag{3.2a}$$

$$(u^*, v^*, w^*) = (u, v, w)/U_c, \; T^* = T/T_{c,} \rho^* = \rho/\rho_c, \; p^* = p/p_c. \tag{3.2b}$$

*In this case*, in the *typical dimensionless NSF initial-boundary value problem*, for a *thermally perfect gas*, *eight* main *non-dimensional reference parameters/numbers emerge*:

$$\begin{aligned}
&St = L_c/U_c t_c \quad \text{Strouhal number},\\
&Re = U_c L_c/\nu_c \quad \text{Reynolds number},\\
&M = U_c/c_c \quad \text{Mach number},\\
&Pr = Cp\ \mu_c/k_c, \quad \text{Prandtl number},\\
&Bo = \gamma(M/Fr_{Lc})^2 \quad \text{Boussinesq number},\\
&\Lambda = \chi/T_w \quad \text{wall temperature parameter},\\
&P\acute{e} = Re\,Pr \quad \text{Péclet number},\\
&Fr_{Lc} = U_c/\sqrt{g L_c} \quad \text{Froude number}.
\end{aligned} \tag{3.2c}$$

In (3.2c), the Mach number M is defined by $c_c = \sqrt{\gamma R T_c}$, which is the characteristic value of the *speed of sound.*

Thanks to (3.2c) the dimensionless NSF equations can be written, for a *thermally perfect gas*, in the following form, when the *Stokes relation* is assumed:

$$\begin{aligned}
&St\,Du/Dt + \left(1/\gamma M^2\right)\nabla p + \left(Bo/\gamma M^2\right)\ \rho \boldsymbol{k}\\
&= (1/Re)\left\{\nabla^2 u + (1/3)\nabla(\nabla.u)\right\},
\end{aligned} \tag{3.3a}$$

$$St\,D\rho/Dt + \rho\nabla.u = 0, \tag{3.3b}$$

$$St\rho DT/Dt + (\gamma\text{-}1)\ p\nabla.u = (1/Pr\,Re)\nabla^2 T + (\gamma\text{-}1)\left[\gamma M^2/Re\right]\Phi, \tag{3.3c}$$

$$p = \rho T, \tag{3.3d}$$

where the *superscripts* "*" *are omitted* for simplicity.

This *non-dimensional approach* is very well discussed (it seems to me, for the first time) by *Lagerstrom* in Chap. 3 of his *long pioneering* article on *laminar flow theory*, which was written in *1953* but *only* published by *F. K. More in 1964* [11] (see the references at the end of the Chap. 1 in this book).

In the framework of an initial-boundary value *(I-BV)* for the above *dimensionless* NSF equations (3.3a–3.3d), one must first write for each particular case two typical dimensionless *boundary* conditions: for the velocity **u** (the *no-slip condition,* because we take into account the *viscosity effect*) and for the temperature T (which must be *identified with a given temperature field* $\Theta$ on a *wall* ($\Gamma$) of a given *moving solid body*, because we consider the *non-adiabatic case*).

In a *very simplified steady case* when in particular the velocity $\mathbf{u}_w$ is such that $|\mathbf{u}_w| \equiv U_c$ and $T_w/T_c = 1$, we can write

$$(u, v, w) = 1, \text{ on a wall, } \Gamma, \text{ of a solid body at rest,} \tag{3.4a}$$

$$T = 1 + \Lambda\ \Theta, \text{ on } \Gamma, \tag{3.4b}$$

and then the *dimensionless initial conditions* for the unknown functions (u), ρ, and T have the very simple form

$$t = 0: \quad (u, v, w) = 0, \quad \rho = 1, \quad T = 1 \text{ and also } p = 1, \tag{3.4c}$$

which correspond to a given *state of rest.* A general unsteady case, indeed *a very important case*, often involving some anxiety, is *the meteo problem*: *what will the weather be like tomorrow or for the next few days*? Indeed, the *initial conditions, given known data* for the previous day, play a particularly crucial role in *weather forecasting*!

For an *external problem some conditions on the behaviour far from the solid body* are also necessary in the various cases, and in an *internal unconfined aerodynamical problem*, when the fluid flow is considered *inside a bounded domain (a cavity* $\Omega$ *with a wall* $\Sigma(t)$ *that can deform in time)*, a very realistic unsteady problem, for the full NSF equations (3.3a–3.3d), was formulated by *Guiraud* and *Zeytounian, 1980 (and see Chap. 5 in [3] cited in our above Chap. 2). The reader is also referred to the discussion of the Mach number below for further information concerning this internal aerodynamical problem.*

In its generality, the typical I-BV dimensionless equations (3.3a–3.3d), with the rather simplified boundary conditions (3.4a and 3.4b), and initial conditions (3.4c), produce an *(I-BV)* NSF problem (discussed in my paper in Russian Math Surveys <1>) that is actually rather complex and difficult to understand, especially in the framework of (pure!) "*mathematical fluid dynamics*"! *However*, this problem turns out to be rather surprising, because the presence of the various *reference parameters* St, Re, $M^2$, Pr, Bo, $\Lambda$, Pé, and $Fr_{Lc}$ in (3.2c) opens the way to *fruitful applications of the RAM approach* via deconstruction!

In the four sections of Chap. 6 and various sections of Chap. 8, the reader can find, in addition to the above list (3.2c) of *non-dimensional parameters, several other parameters arising in the framework of various working models*:

The *Strouhal* number, St, which is related to the unsteadiness of the relevant fluid flow, can take the value $St \ll 1$ (then $t_c \gg U_c/L_c$), which is the case of a *long time evolution* problem, but also $St \gg 1$ (then $t_c \ll U_c/L_c$), which is the case of an *unsteady, transient, short time adjustment towards an evolution problem.* But the case of *high Strouhal number* plays a more interesting role when we also consider $Re \gg 1$, so that $St \to \infty$ and $Re \to \infty$ simultaneously. This is the case of a *high frequency oscillating viscous flow.* In fact, if $T_{osc} \ll L_c/U_c$ is a *period*, and $\omega_{osc} \gg U_c/L_c$ the *high frequency oscillation*, then it makes sense to *introduce the small parameter*

$$a_c = U_c/\omega_{osc}L_c << 1, \tag{3.5a}$$

and consider the limiting process

$$a_c \to 0 \text{ and } Re = U_cL_c/\nu_c \to \infty. \tag{3.5b}$$

This leads to an oscillating *double boundary layer model*! Often, in place of the above, it seems more appropriate to consider

$$a_c \to 0, \ Re \to \infty \text{ and } M \to 0. \tag{3.5c}$$

The theory based on (3.5c) is the so-called *double boundary layer* phenomenon, analysed first by *Stuart (1966)* and *Riley (1965), and expounded in Chap. XI* of [12] cited in our above Introduction. This leads to the *steady acoustic streaming effect* (see also Fig. 3.1).

High frequency oscillations in a fluid initially at rest cause a steady-state secondary flow through the action of viscosity in the boundary layer. This secondary flow (see Fig. 3.1) is such that the entire fluid close to the oscillating body is set into steady-state motion, although the motion of the body is purely periodic. The above calculation is due to *Schlichting (1932)* and shows the streamline portrait of this steady-state additional flow for an oscillating circular cylinder.

In Chap. XI of our 1987 book [12], the reader will find further discussion, e.g., concerning the *Riley-Stuart double boundary layer phenomenon,* and in particular, a detailed asymptotic analysis (pp. 265–278) of a circular cylinder oscillating perpendicularly to generating lines, when the initial working model is that of a Navier incompressible and viscous fluid.

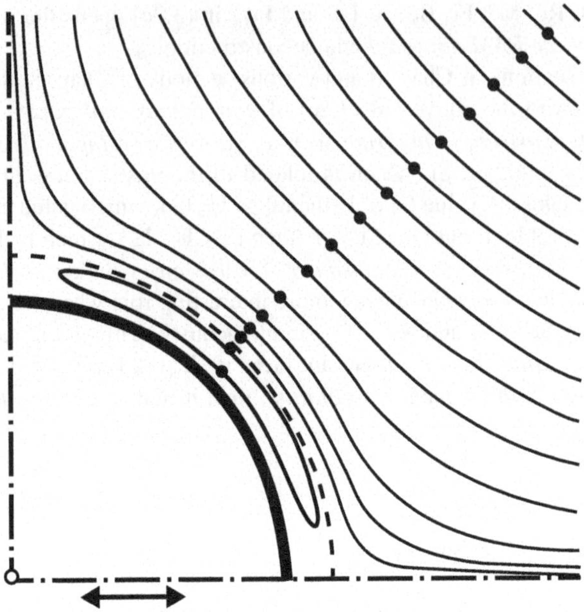

**Fig. 3.1** Streaming close to an oscillating circular cylinder. *From p. 191 of* [14] cited in Introduction

We observe also that the similarity parameter

$$\mathrm{Re}_s = a_c \ \mathrm{Re} \tag{3.5d}$$

plays an important role. In particular, when $\mathrm{Re}_s \gg 1$, the unsteady Stokes layer is close to the wall and submerged by a steady external boundary layer flow; but on the contrary, when $\mathrm{Re}_s \ll 1$, this steady external flow is a Stokes flow and satisfies the following Stokes equations (see Chap. VII in [12]) for the velocity $\mathbf{u}_s$ and reduced pressure perturbation relative to the pressure at infinity, viz., $p - p_\infty = (1/\mathrm{Re})\ \Pi_s$:

$$\nabla.\mathbf{u}_s = 0, \quad \nabla^2\mathbf{u}_s = \nabla\Pi_s. \tag{3.5e}$$

In *Chap. XI* of [12] (pp. 287–296), this *double boundary layer phenomenon* is also *analysed by starting with* the *Prandtl boundary layer incompressible equations.*

The *Reynolds* number, Re, plays a more important role because, with its companion *Prandtl number* Pr, it governs the *process of dissipation* in fluid flows. In a simple-minded way, we can say that $\mathrm{Re} \gg 1$ leads to inviscid (*à la Euler*) fluid flows, while $\mathrm{Re} \ll 1$ leads to strongly viscous (*à la Stokes*) fluid flows. But this is not the end of matter, and in Sect. 6.1, the case of *high Reynolds* number $Re \gg 1$ is discussed *thoroughly*. This will show that, in the *strongly unsteady case*, when $\mathrm{Re} \gg 1$, with $\mathrm{St} = O(1)$ fixed, instead of the usual *Euler-Prandtl*, non-interacting and *regular/hierarchical coupling*, we must consider a more complicated *five region structure, as sketched in* Fig. 6.1.

This new discovery relating to *high Re asymptotics looks interesting*, but it will require *deeper investigations*. Indeed, it could be *an interesting subject for a PhD thesis!*

In contrast, the case of *small* $\mathrm{Re} \ll 1$—unsteady fluid flow with a *high viscosity*—requires one to consider a *three-region* structure. In the concluding remarks to *Part II*, the reader will find *some results for* $\mathrm{Re} \ll 1$—*unsteady fluid flows*, which are *rather poorly understood.*

The *Mach* number, M, plays an important role in its connection with acoustics and *incompressible hyposonic* fluid flows in the framework of the *Navier-Fourier incompressible viscous* and *non-adiabatic model discussed in Chap. 5* of [7] (see in Introduction page XXVI) on internal flows. In Sect. 6.2.1 *and* 6.2.2 of the present book, *and in particular* Fig. 6.2, *the reader will also find various results relating to the limiting process* $M \to 0$.

The case when we consider an *internal unconfined aerodynamics problem—a time-dependent fluid flow inside an (enclosed) cavity whose shape and volume change with time—is highly complicated and involves a multiple-time-scale homogenization method with averaging, as happens also in* Sect. 8.9, *in the asymptotic derivation of the KZK equation of nonlinear acoustics.*

In the paper by Zeytounian and Guiraud <2>, for the BAIL III Short course in Dublin, June 18–19, 1984, devoted to "*Asymptotic features of low Mach number flows in aerodynamics and in the atmosphere*" the reader can find a review of some models of flows which occur at low Mach number M and high Reynolds number Re, where *two rather different situations* are examined. The first concerns conventional aerodynamics and the second atmospheric flows.

Curiously enough the *acoustic model* comes on the scene even in a situation that is apparently governed by incompressible aerodynamics—this is due to non-uniformities, a spatial one *near infinity* in the case of *unconfined flow* and a *temporal one for small times* and for *high frequency* oscillations in the case of *confined flows*. But the most *intriguing features* are found when one deals with low *Mach number flows in the atmosphere*—it is somewhat *discouraging* to realize that the *low Mach number approximation* looks so good at the outset but that useful results have only been obtained with difficulty by using it! For instance, the so called *quasi-non-divergent flow* involves *strange features* tied to the phenomenon of *blocking* well known in stratified flows at *low Froude number*, and this might explain why they have not been used in *weather prediction*. But *blocking is avoided* whenever the low Mach number approximation is coupled with another approximation!

We mention two cases here. When both the *Rossby and the Mach* numbers are *low*, one gets the useful *quasi-geostrophic approximation* due to *Kibel*, 1963 (and see <3>, for the English translation of the Kibel book, edited in Moscow in

Russian in 1957). This has been *systematically investigated, from the point of view of the RAM approach*, by *Zeytounian* and *Guiraud* and *Zeytounian, in 1980 (and see Chap. 11 in* [22] cited in Introduction).

When both the Boussinesq and the Mach number are low (as discussed in Sect. 2.6), *Zeytounian* in *1974* <4> *was the first to show* that one gets a useful Boussinesq model—especially for lee waves—and this led Guiraud and Zeytounian <5>, to a *revisited version* of the so called *Long (1955)* model of lee waves, from the point of view of *asymptotics*.

The *Boussinesq* number, Bo, plays a crucial role in the derivation/justification of the famous *Boussinesq approximate model equations*, as was shown in Zeytounian <4>. It was associated with the Mach number by the similarity relation

$$\mathrm{Bo}/\mathrm{M} = \gamma \mathrm{M}/(\mathrm{Fr_c})^2 \equiv \mathrm{B}^* = \mathrm{O}(1), \tag{3.6a}$$

which shows the role of the Froude number $\mathrm{Fr_c}$ [see (3.2c)], where

$$\mathrm{L_c} = \mathrm{H_c} = \mathrm{RT_c}/g \approx 10^4 \ \mathrm{m} \tag{3.6b}$$

is taken as the height of a homogeneous hydrostatic (standard) atmosphere. But if we take into account the usual physical values for atmospheric motions, then from (3.6b) we find that the *Boussinesq layer* is only of the order $10^3\ \mathrm{m} \ll \mathrm{H_c} \approx 10^4\ \mathrm{m}$, and this is a *strong restriction* on the application of the Boussinesq equations to atmospheric motions.

The prediction of "*steady lee waves*" in the whole troposphere around and downstream of a mountain is considered in Sect. 8.6. The reader will find *valuable information* about the *justification* and *model equations emerging* from Newtonian fluid dynamics for weather forecasting in the paper Zeytounian <8>, cited in Chap. 2, which is the *English translation* of our paper in Russian, and in [22], cited in "Introduction", <2>, cited in Chap. 2, Chap. 11 of [7], cited in "Introduction", and Chap. 6 of [24], cited in "Introduction".

In fact, *meteo-fluid dynamics* is a privileged discipline for the application of the RAM approach, because in addition to the Bo, M, and Re numbers, we also have (in the *atmospheric NSF equations*) the *Rossby, Kibel, Ekman* (Ro, Ki, Ek) numbers and the *hydrostatic parameter* ε, i.e., the ratio between the vertical and horizontal length scales chosen for atmospheric motion, as was shown in Sect. 2.6.

*In* Sect. 3.4, *I have decided to show the case of dimensionless equations for atmospheric motions, when the latter equations are written in spherical coordinates (!). In the framework of the application of the RAM approach, these reveal a useful family of reduced approximate working models.* A sketch of these approximate *meteo-fluid dynamics models appears* in Fig. 6.3, *in* Sect. 6.3. These working meteo models were derived from the full unsteady *non-dimensional atmospheric NSF* equations during the years 1976–1991.

Let me mention in particular the book *Zeytounian, 1990* [22], a *translation from French to English* of my *hand-written* manuscript "*Modelisation Asymptotique des Ecoulements Atmosphériques*", 596 pages, University of Lille-I, 1985, in which meteorology is viewed as a discipline of fluid dynamics, and singular perturbation methods using the RAM approach are the main tools. It seems to me that (*unfortunately!*) the main "defect" of this monograph is the very *poor English translation*, despite the very positive critical review by *Huijun Yang* (University of Chicago).

*My 1991 book Meteorological Fluid Dynamics* <3>, cited in Chap. 2, should be a good preparation for reading the above mentioned monograph (1990)! In my "scientific autobiography" published in 2014, the second volume of my "*trilogy*", the reader will find a *short critical review* of this 1991 book, which appeared in *Appl Mech Rev 46(42) B29*, 1993, with the following rather *positive conclusion* by *Fred R. Payne* from the Aerospace Engineering Department, University of Texas:

> This is a simulating book—most first-time readers will likely make copious comments upon the margins. It is a member of a select subset, needful of study by specialists in fluid mechanics, turbulence, atmospheric dynamics and modeling, and finite-dimensional dynamic systems.

The *Bénard thermal convection problem*, for a *weakly expansible liquid heated from below*, is considered in Sect. 6.4, *and* some information is given concerning the *three main leading-order working models (*already mentioned above). We show *how the four main driving effects—the buoyancy, the temperature-dependent surface tension, the viscous dissipation, and the deformable free surface—are taken into account—as illustrated in* Fig. 6.4. The key here is to rewrite *the Grashof number Gr as a ratio of two parameters*, namely:

$$\mathrm{Gr} = \alpha/(\mathrm{Fr_d})^2, \tag{3.7}$$

where α is the *dilatation/expansion small parameter, and* $\mathrm{Fr}_d$ the *Froude number* for a *layer* of thickness *d* (see Sect. 6.4).

We observe that such rewritings of various classical parameters as ratios of two other parameters in various mathematically formulated fluid dynamics problems leads to the possible use of the RAM/deconstruction approach, when (as in the case of (3.7), for $\alpha_c$, assumed small) it is assumed that at least one parameter in these ratios is a small (or large) parameter!

The above, seemingly naïve, statement often provides a way to extract reduced working models from very stiff NSF problems—modelling various fluid dynamic flows with a dominant small or large physical effect!

## 3.3 Some Key Steps in the RAM Approach. The Deconstruction Process

Here we should mention a most significant point in the application of the RAM approach. With (3.2c) we already get, in the case of aerodynamics, *heat parameters* from the dimensionless writing of the NSF equations and conditions, and numerous other parameters are considered in the various models relating to liquids, the atmosphere, thermal convection, nonlinear acoustics, and aerodynamics—see for instance the various applications of the RAM approach in Chaps. 6 and 8.

First we note that such a dimensionless writing of the NSF problems requires *some knowledge* of Newtonian fluid dynamics, and also that the flow problem studied does not involve turbulence! Before applying the RAM approach to, *dimensionless fluid dynamics problems*, some knowledge of the *physical features* of the specific flow structure is necessary. Here the 2001 book <6> by *Guyon, Hulin, Petit,* and *Mitescu (Physical Hydrodynamics, Oxford University Press) is highly valuable, being mainly based on the lectures by Pierre-Gilles De Gennes (Nobel Prize for Physics, 1991, de Gennes was director of the Ecole supérieure de physique et de chimie industrielle de la ville de Paris from 1976 to 2002).*

In the RAM approach, some of the parameters are taken, for the case at hand, to be close to (or *to tend* to) zero or infinity! Indeed, we have to go farther (as in the case considered with Godts in 1990), since *Re is close to infinity while M is close to zero. The case considered below in* (3.8a) gives very valuable results when (3.8b) is satisfied!

*The key steps which are often used are of great value in the RAM approach. They are related to the discovery of various similarity rules, which are essential here,* because in many cases one encounters a double or multiple limiting process, in which two or more reference parameters approach their limits. *One must frequently specify the relative rates in which the limits are approached, since the order in which they are carried out cannot in general be interchanged.*

As a typical example I choose here the case I considered with *Godts* in 1990, when $\mathrm{Re}\uparrow\infty$ and $\mathrm{M}\downarrow 0$ *simultaneously*, such that

$$1/\mathrm{Re} = \kappa_0 \mathrm{M}^{\beta}, \text{ with } \beta > 0 \text{ and } \kappa_0 = O(1), \tag{3.8a}$$

where $\kappa_0$ is a fixed scalar similarity parameter of O(1) which must be determined during the RAM approach, and is selected during the matching.

The above case (3.8a) is the most significant and leads to a *degenerate leading-order working model.* For instance, for the classical *Blasius problem* for a *slightly compressible flow* with vanishing viscosity, a consistent RAM approach leads to (see Sect. 8.5)

$$\beta = 4 \text{ with}: \ \kappa_0 \approx 1 \Rightarrow (1/\mathrm{Re})/\mathrm{M}^4 \approx 1. \tag{3.8b}$$

We observe that, *when we assume* $\kappa_0\uparrow\infty$, we rediscover only a *very degenerate model* corresponding to

$$\mathrm{Re} \text{ fixed}, \ \mathrm{M}\downarrow 0, \text{ and then } \mathrm{Re}\uparrow\infty, \tag{3.8c}$$

and *when* $\kappa_0\downarrow 0$, we rediscover again a *very degenerate model* corresponding to

$$\text{M fixed, Re} \uparrow \infty, \text{ and then M} \downarrow 0. \tag{3.8d}$$

Concerning the *matching*, which is a *crucial step in the application of the RAM approach, and in the book by Van Dyke, 1975* [17] cited in Introduction, the reader can find a detailed consideration of various facets of this matching. In particular, to *leading order*, when we consider two limiting processes in two contiguous regions, as is the case for $Re\uparrow\infty$, we derive *two systems of equations* from the NSF system of equations: First, the *Euler non-viscous and adiabatic equations, valid in the region outside the boundary layer close to the wall*, and then the *Prandtl boundary-layer, viscous and non-adiabatic equations close to the wall of a body in motion. These two systems of equations are* derived, respectively, via the following *two limiting processes, outer (Euler) and inner (Prandtl), when* $\varepsilon = 1/\surd\text{Re} \downarrow 0$:

$$\text{Lim}^{\text{E}}(t \text{ and } x, y, z \text{ fixed}) \text{ and } \text{Lim}^{\text{Pr}}(\text{tand } x,\ y,\ \zeta = \text{z}/\varepsilon \text{ fixed}). \tag{3.9a, b}$$

A *simplified matching condition is written to leading order* in the following form:

$$\begin{aligned} &lim_{\zeta\uparrow\infty}\{\text{Lim}^{\text{Pr}}(\text{t and } x,\ y,\ \zeta = \text{z}/\varepsilon \text{ fixed}) \\ &\quad = lim_{\text{z}\downarrow\text{o}}\{\text{Lim}^{\text{E}}(\text{t and } x,\ y,\ z \text{ fixed})\}. \end{aligned} \tag{3.9c}$$

As a matter of fact, in various cases, an *intermediate "overlap" region* often *exists*, where both the *inner* and *outer* limiting processes are valid, and the matching is realised via the *intermediate (matching) equations*. This is the case in Sect. 6.1, *where the corresponding intermediate/matching equations are derived!*

Finally, a *crucial step* in the *MAE* is the choice of inner variables; the main guiding principle is that the *leading-order inner problem* should have the *least possible degeneracy* and include, in the leading-order working model, any essential *elements omitted* in the leading-order *outer problem*, in such a way that *the two problems should match*. On the other hand, it should be noted that various problems in fluid dynamics are characterised by the presence of *small disturbances* which, because they are *active over a long time* on some portion of the flow, have a non-negligible cumulated effect! In such a case, matching is impossible and one must work with the *method of multiple scales (MMS)*.

This *MMS* is indeed a particular case of the general *homogenization technique*. In this case, *the first step is typically the identification of micro and macro scales*. In 1980, the *limiting case* $M\downarrow 0$ *was considered* with Guiraud for an *internal aerodynamics problem*, namely, the *time dependent fluid flow inside an (enclosed) cavity whose shape and volume change with time*. Just such a *homogenization was* considered, and the reader can find this application in Chap. 5 of [24] cited in Introduction. The most important result we obtained with Guiraud is the following:

> If the motion of the wall of the deformable (as a function of time) cavity, in which an inviscid gas is confined, is started impulsively from rest, then acoustic oscillations remain present, and have an effect on the pressure which would be felt by a gauge, and are not

related to the mean (averaged) motion. The same holds if the wall is accelerated from rest to a finite velocity in a time O(M). In particular, in a closed piston-cylinder system, the isentropic compression due to slow motion is modified by acoustic waves!

It is interesting to note also that, in fact earlier, *in 1970, Guiraud* and I used such a *homogenization technique (maybe for the first time ?) without realising it,* in [19] and [20] cited in Introduction, which are devoted to the *modeling of turbomachinery fluid flow*, when the *long scale* is clearly the one related to the *mean* (*macro*) *flow in the whole turbine*, and the *short* (*micro*) *scale* is the one related to the *thin space* between two individual blades.

The *obvious small parameter is* 1/*N, where N is the number of blades in a row* or the number *of stages of rows. The source term (which is proportional to the jump in pressure) in the momentum equation for the derived (with an error of* $(1/N)^2$*) axially symmetrical through-flow occurs from redistribution (homogenization!) of forces exerted on the flow by the blades. In other words, one finds that the momentum equation for the axial momentum of the mean homogenized through-flow over the global turbine involves a force exerted by the blades which may be estimated to a leading approximation (*see Sect. 8.2*).*

*If we are to obtain approximate reduced working models at least to leading order with our RAM approach, it is absolutely necessary to start with the full dimensionless NSF system of equations and take into account the initial and boundary conditions, without any simplifications! In* Sect. 3.4, *the reader will see that such a requirement is not always easy to fulfill! Unfortunately, it is often recommended, somewhat paradoxically, to choose, during the step to dimensionless parameters, the simplest case, neglecting* a priori *various seemingly (!) unimportant terms in the starting equations; and only sometimes to consider, when is possible, a more complicated case!* This popular approach is for me the obvious way to *cast aside certain valuable approximate reduced working models, and to obtain non-rational and rather meaningless models instead.*

In fact, *our logical motto is always start with the more complicated case, and only subsequently consider particular cases, taking into account the RAM approach by focusing on small or large parameters in dimensionless problems!* As an example, I mention here the so-called full *hydrostatic viscous and non-adiabatic, non-tangent system of equations.* These equations are derived in Sect. 3.4 from the *full unsteady NSF atmospheric equations written in spherical coordinates linked with the rotating Earth, and they involve two effective small parameters, namely* $\varepsilon_0 = H_c/L_c$ and $Re^{-1} = \nu_c/U_c L_c$.

Despite the *smallness* of these two parameters, we have been able to take into account, in the derived working model, not only viscous and heat conduction, but also the influence of the disparate length scales, by assuming that

$$\varepsilon_0 \to 0, \text{ and } Re \to \infty, \tag{3.10a}$$

in such a way that

$$\varepsilon_0^2 \mathrm{Re} \equiv Re_\perp = \mathrm{O}(1), \quad \text{and } t', x', y', z' \text{ fixed}, \tag{3.10b}$$

*in order to get a generalisation of the well known Kibel primitive equations (discovered in 1957;* see, for instance, Sect. 9.2 of our 2012 companion book cited as [5] in references of Preface).

The above similarity relation (3.10b), with $\mathrm{Re}_\perp = \mathrm{O}(1)$, plays a crucial role in the case of weakly dissipative effects—*the coupling of two parameters*, respectively *small* and *high*, makes it necessary to *take into account* the *viscous* and *heat conducting terms* in the *viscous and non-adiabatic, non-tangent system of equations, which generalize the well-known Kibel primitive equations, despite the fact that the terms proportional to* $\varepsilon_0$ *and Re considered separately are negligible in the full starting NSF equations!*

*My first experience (see, for instance,* Sect. 2.6*) with the justification and derivation of the Boussinesq reduced equations from the NSF equations in 1954 gives me confidence that the following rather difficult derivation of various working models for atmospheric motions will justify the above opinion!*

## 3.4 The Case of Dimensionless Equations: In Spherical Coordinates—For Atmospheric Motions and Some Reduced Approximate Working Models

In Sect. 2.6, *the reader had a taste of the particular effects of a rotating earth on fluid motion (assuming dry atmospheric air)*. These effects clearly show that so-called *geophysical fluid dynamics* poses a multitude of interesting questions in mathematical physics. Curiously, some of them are also relevant to *magneto-hydrodynamics*, because of the analogy between the *Coriolis force* and the *Lorentz force* $\mathbf{H} \times \mathbf{u}$, on an *ionized fluid* under a *magnetic* field **H** (see, for instance, Sect. 10.3 in my book [25] published in 2009 and cited in Introduction).

In a coordinate frame *rotating with the earth*, the *momentum* and the *continuity equation are*

$$\rho D\mathbf{u}/Dt = \rho\mathbf{g} - \nabla p - \rho(2\boldsymbol{\Omega} \times \mathbf{u}) + \mu_O \nabla^2 \mathbf{u} + (\mu_O/3)\nabla(\nabla \cdot \mathbf{u}), \tag{3.11a}$$

$$D\rho/Dt + \rho\nabla.\mathbf{u} = 0, \tag{3.11b}$$

where **u** is the relative velocity and $\rho$ the density. As *energy equation*, for the specific internal energy e, we have

$$\rho De/Dt + (p/\rho)D(1/\rho)/Dt = k_O \nabla^2 T + (\nabla\mathbf{u}).\ \underline{\boldsymbol{T}} + \rho_{st} Q_{st}(T_{st}), \tag{3.11c}$$

according to (2.14c). In equations (3.11a, 3.11b) and (3.11c), the dissipative coefficients $\mu_O$ and $k_O$ are assumed constant, and the term $(\underline{\nabla\mathbf{u}}) \cdot \underline{\boldsymbol{T}}$, where $(\underline{\nabla\mathbf{u}})$

is a second-order tensor and $\underline{T}$ the viscous stress, a second-order tensor, is expressible as

$$\begin{aligned}(\underline{\nabla \mathbf{u}}).\underline{T} &= \mu_O\{\underline{\nabla \mathbf{u}} \,.\, def\,\mathbf{u} - (2/3)(\nabla.\mathbf{u})^2\} \\ &= \mu_O\Big\{-(2/3)(\nabla.\mathbf{u})^2 + (½)\,[(def\,\mathbf{u})\,.\,(def\,\mathbf{u})]\Big\},\end{aligned} \tag{3.12a}$$

in which $def\,\mathbf{u} = (\underline{\nabla \mathbf{u}}) + (\underline{\nabla \mathbf{u}})^*$ is the symmetric velocity deformation tensor. For the acceleration Du/Dt, a short calculation shows that

$$\mathrm{D}\mathbf{u}/\mathrm{Dt} = \partial\mathbf{u}/\partial t + (\mathbf{u}.\nabla)\mathbf{u} = \partial\mathbf{u}/\partial t + \nabla\left(q^2/2\right) + 2\boldsymbol{\omega}\times\mathbf{u}, \tag{3.12b}$$

where $q = |\mathbf{u}|$ and $\boldsymbol{\omega} = (½)\,(\nabla \times \mathbf{u})$. Just from this form, the components of Du/Dt are easily found for *any system of curvilinear orthogonal coordinates*.

Finally, with the *equation of state* for the *pressure of dry air* and the associated thermodynamic relation between the specific internal energy e and the absolute temperature T, viz.,

$$p = R\rho T \text{ and } e = Cv\,T, \tag{3.13a, b}$$

the above equations (3.11a), (3.11b), (3.11c), with (3.12a), (3.12b) and (3.13a–b) provide a closed system for **u**, ρ, p, e, and T.

Now it is *helpful to use spherical coordinates* λ, φ, r, and let u, v, w denote the corresponding components of the velocity **u** in the direction of increasing azimuth (λ), latitude (φ), and radius (r) respectively. In terms of these coordinates the metric is given by

$$(ds)^2 = r^2\cos^2\varphi\,(d\lambda)^2 + r^2(d\varphi)^2 + (dr)^2. \tag{3.13c}$$

Hence we get:

$$\nabla = (1/r\cos\varphi)\,(\partial/\partial\lambda)\mathbf{i} + (1/r)\,(\partial/\partial\varphi)\mathbf{j} + (\partial/\partial r)\mathbf{k}, \tag{3.13d}$$

and in Sect. 4 of Chap. II of our 1991 book, with the title *Meteorological Fluid Dynamics*, the reader can find the various formulas needed to rewrite the atmospheric NSF equations in these spherical coordinates.

As is well observed by *Fred R. Payne* in the above-mentioned review of the book *Meteorological Fluid Dynamics*):

> … it is apparent that Zeytounian has expended a considerable effort to make the material in this 1991 book accessible to non-specialists.
>
> This is especially the case in the derivation of the full system of dimensionless atmospheric NSF equations (3.14a)–(3.14f) with (3.15).

First, in the starting equations (3.11a–3.11c), with (3.12a–3.12c) and (3.13d), we have introduced the transformation (3.14), and the origin of this new right-handed curvilinear coordinate system x, y, and z lies on the earth's surface (*flat ground*, $r = a_0$) at latitude $\varphi_0$ and for some specific longitude origin $\lambda_0 = 0$. We suppose therefore that the atmospheric motion occurs in a mid-latitude region, distant from the equator, around some central latitude $\varphi_0$, and assume that $\sin\varphi_0$, $\cos\varphi_0$, and $\tan\varphi_0$ are all of order unity:

$$x = a_0 \cos\varphi_0 \, \lambda, \; y = a_0 \, (\varphi - \varphi_0), \; z = r - a_0. \tag{3.14}$$

Although x and y are in principle new longitude and latitude coordinates, in terms of which the above basic starting equations (3.11a–3.11c) may be rewritten without approximation, they are obviously introduced in the expectation that, for a *small sphericity parameter*

$$\delta_0 = L_0/a_0 << 1, \tag{3.15}$$

they will be the Cartesian coordinates of the so-called *β-plane approximation*!

### 3.4.1 The NSF Generalized Atmospheric System of Equations

As a *dimensionless continuity equation*, we derive first

$$\begin{aligned} &\mathrm{St}\, D'\rho'/Dt' + \rho' \{(\eta_0/\varepsilon_0)\partial w'/\partial z' + 2\delta_0[\eta_0/(1 + \varepsilon_0\delta_0 z')]w' \\ &+[(1/\cos\varphi)/(1 + \varepsilon_0\delta_0 z')]\, \partial(\cos\varphi\, v')/\partial y' \\ &+[(\cos\varphi_0/\cos\varphi)/(1 + \varepsilon_0\delta_0 z')]\, \partial u'/\partial x'\} = 0. \end{aligned} \tag{3.16a}$$

In (3.16a), the four non-dimensional ratios are

$$\eta_0 = W_0/U_0, \;\; \mathrm{St} = L_0/U_0 t_0, \;\; \varepsilon_0 = H_0/L_0, \text{ and } \delta_0 = L_0/a_0,$$

where $a_0$ is the *earth radius*, and convenient scales $W_0$ $U_0$ for the velocities, $H_0$, $L_0$ for the lengths, and $t_0$ for the time have been introduced.

The *good* case is $\eta_0/\varepsilon_0 \equiv 1$ and $W_0 \equiv \varepsilon_0 U_0$, and using this good choice of scales, the *dimensionless azimuthal component* u' *of the above momentum equation* (3.11a) *becomes*

$$
\begin{aligned}
&\rho'\{\mathrm{St}\,D'u'/Dt' + \varepsilon_0[\delta_0/(1+\varepsilon_0\delta_0 z')]u'w' - \delta_0[\tan\varphi/(1+\varepsilon_0\delta_0 z')]u'v' \\
&\quad -[1/\mathrm{Ro}](\sin\varphi/\sin\varphi_0)v' + [\varepsilon_0/\mathrm{Ro}](\cos\varphi/\sin\varphi_0)w'\} \\
&\quad +[1/\gamma M^2][(\cos\varphi_0/\cos\varphi)/(1+\varepsilon_0\delta_0 z')]\partial p'/\partial x' \\
&= (1/\mathrm{Re})\{\Delta' u' + (1/3)[(\cos\varphi_0/\cos\varphi)/(1+\varepsilon_0\delta_0 z')]\partial(\nabla'.\mathbf{u}')/\partial x' \\
&- \delta_0^2\left[(1/\cos^2\varphi)/(1+\varepsilon_0\delta_0 z')^2\right]u' + 2\delta_0\varepsilon_0\left[(\cos\varphi_0/\cos\varphi)/(1+\varepsilon_0\delta_0 z')^2\right]\partial w'/\partial x' \\
&\quad -2\delta_0\tan\varphi\left[(\cos\varphi_0/\cos\varphi)/(1+\varepsilon_0\delta_0 z')^2\right]\partial v'/\partial x'\},
\end{aligned}
\tag{3.16b}
$$

where

$$
\begin{aligned}
\Delta' = &\left[(\cos^2\varphi_0/\cos^2\varphi)/(1+\varepsilon_0\delta_0 z')^2\right]\partial^2/\partial x'^2 + \left[1/(1+\varepsilon_0\delta_0 z')^2\right]\partial^2/\partial y'^2 \\
&+ (1/\varepsilon_0^2)\partial^2/\partial z'^2 - \delta_0\left[\tan\varphi/(1+\varepsilon_0\delta_0 z')^2\right]\partial/\partial y' \\
&+ 2\,[\delta_0/\varepsilon_0][1/(1+\varepsilon_0\delta_0 z')]\partial/\partial z',
\end{aligned}
$$

and

$$
\begin{aligned}
\nabla'.\mathbf{u}' = \; &\partial w'/\partial z' + 2\varepsilon_0\delta_0[1/(1+\varepsilon_0\delta_0 z')]w' \\
&+[(1/\cos\varphi)/(1+\varepsilon_0\delta_0 z')]\,\partial(\cos\varphi\, v')/\partial y' \\
&+[(\cos\varphi_0/\cos\varphi)/(1+\varepsilon_0\delta_0 z')]\,\partial u'/\partial x'.
\end{aligned}
$$

Analogous momentum equations can be written down for v' and the vertical component of the velocity w':

$$
\begin{aligned}
&\rho'\{\mathrm{St}\,D'v'/Dt' + \varepsilon_0[\delta_0/(1+\varepsilon_0\delta_0 z')]v'w' + \delta_0[\tan\varphi/(1+\varepsilon_0\delta_0 z')]u'^2 \\
&+[1/\mathrm{Ro}](\sin\varphi/\sin\varphi_0)u'\} + [1/\gamma M^2][(1/(1+\varepsilon_0\delta_0 z')]\partial p'/\partial y' \\
&= (1/\mathrm{Re})\{\Delta' v' + (1/3)[1/(1+\varepsilon_0\delta_0 z')](\partial(\nabla'.u')/\partial y') \\
&- \delta_0^2[(1/\cos^2\varphi)/(1+\varepsilon_0\delta_0 z')^2]v' + 2\delta_0\left[\varepsilon_0/(1+\varepsilon_0\delta_0 z')^2\right]\partial w'/\partial y' \\
&- 2\delta_0\tan\varphi\left[(\cos\varphi_0/\cos\varphi)/(1+\varepsilon_0\delta_0 z')^2\right]\partial u'/\partial x'\},
\end{aligned}
\tag{3.16c}
$$

and

$$
\begin{aligned}
&\rho'\{\mathrm{St}\,\varepsilon_0^2 D'w'/Dt' - \varepsilon_0\delta_0[1/(1+\varepsilon_0\delta_0 z')](u'^2+v'^2) \\
&-[\varepsilon_0/\mathrm{Re}](\cos\varphi/\cos\varphi_0)u'\} + [1/\gamma M^2](\partial p'/\partial z' + \mathrm{Bo}\rho') \\
&= (1/\mathrm{Re})\{\varepsilon_0^2\Delta' w' - 2\delta_0^2\varepsilon_0^2\left[1/(1+\varepsilon_0\delta_0 z')^2\right]w' - 2\delta_0\varepsilon_0\left[1/(1+\varepsilon_0\delta_0 z')^2\right](\partial v'/\partial y') \\
&+2\delta_0^2\varepsilon_0\tan\varphi\left[1/(1+\varepsilon_0\delta_0 z')^2\right]v' - 2\delta_0\varepsilon_0\left[(\cos\varphi_0/\cos\varphi)/(1+\varepsilon_0\delta_0 z')^2\right]\partial u'/\partial x'\}.
\end{aligned}
\tag{3.16d}
$$

The dimensionless form of the equation of state (3.13a) is simply, for dry atmospheric air and assuming a perfect gas,

$$p' = \rho' T'. \tag{3.16e}$$

Finally, instead of (3.10c) with (3.12b), we obtain for the dimensionless temperature $T'$ the following dimensionless equation of energy:

$$\begin{aligned} &\rho' \mathrm{St}\, D'T'/Dt' - [(\gamma-1)/\gamma]\mathrm{St}\, D'p'/Dt' = (1/\mathrm{Pr\,Re})\ \Delta' T' \\ &+\left[(\gamma-1)/{\varepsilon_0}^2\ \mathrm{Re}\right]M^2\ \chi'(\mathbf{u}',\ \varepsilon_0,\ \delta_0) \\ &+(1/\mathrm{Pr})\ \left[\mathrm{Bo}^2\sigma_0/{\varepsilon_0}^2\mathrm{Re}\ d\left[R_{st}(T_{st})\right]/dz_{st}, \end{aligned} \tag{3.16f}$$

where $\sigma_0 = R/gk_0$, and $\mathrm{Pr} = \mu_0 Cp/k_0$ is the *Prandtl number*.

For the function $\chi'$—*the viscous dissipation function in* (3.16f)—we obtain the following rather awkward formula (3.17):

$$\begin{aligned} \chi' = {\varepsilon_0}^2\left[1/(1+\varepsilon_0\delta_0 z')^2\right]&\{\{(2/3)\{\left[(\cos\varphi_0/\cos\varphi)\partial u'/\partial x' - \delta_0\tan\varphi\, v' - \partial v'/\partial y'\right]^2 \\ &+[\partial v'/\partial y' + \varepsilon_0\delta_0 w' - (1+\varepsilon_0\delta_0 z')(\partial w'/\partial z')]^2 + \left[(1+\varepsilon_0\delta_0 z')\partial w'/\partial z'\right. \\ &\left.-(\cos\varphi_0/\cos\varphi)\partial u'/\partial x' + \delta_0\tan\varphi\, v' - \delta_0\varepsilon_0 w'\right]^2]\} \\ &+[\cos\varphi\,\partial(u'/\cos\varphi)/\partial y' + (\cos\varphi_0/\cos\varphi)(\partial v'/\partial x')]^2 \\ &+\left[(1/\varepsilon_0)(1+\varepsilon_0\delta_0 z')^2\partial(v'/[1+\varepsilon_0\delta_0 z'])/\partial z' + \varepsilon_0\partial w'/\partial y'\right]^2 \\ &+\left[\varepsilon_0(\cos\varphi_0/\cos\varphi)\partial w'/\partial x' + (1/\varepsilon_0)(1+\varepsilon_0\delta_0 z')^2\partial[u'/[1+\varepsilon_0\delta_0 z']]^2/\partial z'\}\}. \end{aligned} \tag{3.17}$$

the reader can find the above formula in my book edited in 1991 as <Meteo-fluid dynamics>. LNPm5 by springer edited in Cha. 2 as <3>.

### 3.4.2 Hydrostatic Equations Emerging from the Limiting Process (3.10a and 3.10b) with $\delta_0$ Fixed

In any realistic amospheric situation, $\mathrm{Re} \gg 1$, and if we consider the main hydrostatic viscous and heat conducting limiting process (3.10a and 3.10b), *assuming* $\delta_0 = O(1)$, then in such a case we obviously have $\varepsilon_0 \ll 1$ in the troposphere.

From the full equations, (3.16a)–(3.16f) written in spherical coordinates, a set of dimensionless "*hydrostatic equations*" for *large-synoptic scale atmospheric processes is then derived when we take into account (3.10) and assume that* $\delta_0$ is fixed. In particular, *dropping all the primes,* we obtain the following equations (3.18a)–(3.18e), which are reduced *hydrostatic equations for the non-tangent, viscous and non-adiabatic, large-synoptic scale atmospheric motions:*

$$\mathrm{St}\,D\rho/Dt + \rho\{\partial w/\partial z + (1/\cos\varphi)/\partial(\cos\varphi\, v)/\partial y$$
$$+(\cos\varphi_0/\cos\varphi)\,\partial u/\partial x\} = 0, \tag{3.18a}$$

$$\rho\{\mathrm{St}\,Du/Dt - \delta_0 \tan\varphi u\, v - (1/\mathrm{Ro})(\sin\varphi/\sin\varphi_0)\, v\}$$
$$+ (\cos\varphi_0/\cos\varphi)\left[1/\gamma M^2\right]\partial p/\partial x = (1/\mathrm{Re}_\perp)\partial^2 u/\partial z^2, \tag{3.18b}$$

$$\rho\left\{\mathrm{St}\,Dv/Dt + \delta_0 \tan\varphi\, u^2 + [1/\mathrm{Ro}]\,(\sin\varphi/\sin\varphi_0)\, u\right\} + \left[1/\gamma M^2\right]\partial p/\partial y$$
$$= (1/\mathrm{Re}_\perp)\partial^2 v/\partial z^2, \tag{3.18c}$$

$$\partial p/\partial z + \mathrm{Bo}\ \rho = 0, \tag{3.18d}$$

$$\rho \mathrm{St}\,DT/Dt - [(\gamma-1)/\gamma]\mathrm{St}\,Dp/Dt = (1/\mathrm{Pr})\partial^2 T/\partial z^2$$
$$+[(\gamma-1)/\mathrm{Re}_\perp]M^2\left[(\partial u/\partial z)^2 + (\partial v/\partial z)^2\right]$$
$$+(1/\mathrm{PrRe}_\perp)\,\mathrm{Bo}^2\sigma_0 d[R_{st}(T_{st})]/dz_{st}, \tag{3.18e}$$

$$p = \rho^{T,} \tag{3.18f}$$

where $\mathrm{St}\,D/Dt = \mathrm{St}\,\partial/\partial t + (\cos\varphi_0/\cos\varphi)\,u\partial/\partial x + v\partial/\partial y + w\partial/\partial z,$

and $\mathrm{Re}_\perp = \varepsilon_0{}^2\mathrm{Re}$.

The above *hydrostatic model equations* (3.18a)–(3.18e) constitute a *very significant* system for *large-synoptic scale* atmospheric motions in the *troposphere—a thin layer around the earth—despite the fact that the term Dw/Dt is absent in* (3.18d)—and instead of the above equation (3.16d) for w', we have only a very strongly reduced equation (3.18d), where curiously the vertical component of the velocity w is absent!

### 3.4.3 Kibel Primitive Equations

When $\mathrm{Re}_\perp \equiv \infty$, we obtain, instead of the full system of equations (3.18a)–(3.18e), the set of equations for *non-tangent* and *non-viscous adiabatic atmospheric motions*. The case considered below *with* $\delta_0 \to 0$ is a slight generalization of the famous *Kibel (1957/1963)* "*primitive*" equations, as they are known <3>. The *hydrostatic character* of equations (3.18a)–(3.18e) is mainly due to the *considerable simplification* of the equation for the vertical component of the velocity w (relative to z)!

Concerning the *Kibel primitive equations* for the tangent, hydrostatic non-viscous *and* adiabatic *atmospheric motions*, these are derived from the above system of equations (3.18a)–(3.18f) when $\delta_0 \to 0$, with the relation (*β-effect*):

$$\beta = \delta_0/\mathrm{Ro}\ \tan\varphi_0 \approx 1, \tag{3.19}$$

and they can be written in the following form:

$$\mathrm{St}\ \mathrm{D}\rho/\mathrm{Dt} + \rho\ (\partial u/\partial x + \partial v/\partial y + \partial w/\partial z) = 0, \tag{3.19a}$$

$$\rho\{\mathrm{St}\ \mathrm{Du}/\mathrm{Dt} - [(1/\mathrm{Ro}) + \beta y]v\} + \left[1/\gamma M^2\right]\ \partial p/\partial x = 0, \tag{3.19b}$$

$$\{\mathrm{St}\ \mathrm{Dv}/\mathrm{Dt} + [(1/\mathrm{Ro}) + \beta y]\ u\} + \left[1/\gamma M^2\right]\ \partial p/\partial y = 0, \tag{3.19c}$$

$$(1/\rho)\ \partial p/\partial z + \mathrm{Bo} = 0, \tag{3.19d}$$

$$p = \rho T, \tag{3.19e}$$

$$\rho \mathrm{St}\ \mathrm{DT}/\mathrm{D}t - [(\gamma - 1)/\gamma]\mathrm{St}\ \mathrm{Dp}/\mathrm{Dt}\ = 0, \tag{3.19f}$$

where

$$\mathrm{St}\ \mathrm{D}/\mathrm{Dt} = \ \mathrm{St}\ \partial/\partial t + \ u\partial/\partial x\ + v\partial/\partial y\ + w\partial/\partial z.$$

It is obvious, *and this has a crucial importance when considering the "transient adjustment problem" for the above hydrostatic balance* (3.19d), that the above *Kibel primitive equations* (3.19a)–(3.19f) *can also be obtained through the limiting process*

$$\varepsilon_0 \to 0, \text{ keeping time t and coordinates x, y, and z fixed,} \tag{3.20}$$

*applied to the following full equations* for *tangent, non-hydrostatic, non-viscous, and* adiabatic *atmospheric motions*. Hence, for $\mathbf{v} = (u, v)$, w, ρ, p, and T, we have the following full system of Eulerian equations:

$$\mathrm{St}\ \mathrm{D}\rho/\mathrm{Dt} + \rho(\mathbf{D}.\ \mathbf{v} + \partial w/\partial z) = 0, \tag{3.21a}$$

$$\rho\left\{\mathrm{St}\ \mathrm{D}\mathbf{v}/\mathrm{Dt} - [(1/\mathrm{Ro}) + \beta y](\mathbf{k} \times \mathbf{v}) + \left[(\varepsilon_0/\tan\varphi_0 \mathrm{Ro}\right]\ w\ \mathbf{i}\right\}$$

$$+\left[1/\gamma M^2\right]\mathrm{Dp} = 0, \tag{3.21b}$$

$$\rho\left\{\mathrm{St}\,\varepsilon_0{}^2 \mathrm{Dw}/\mathrm{Dt}\ - \left[(\varepsilon_0/\tan\varphi_0 \mathrm{Ro}\right]\mathbf{v}.\mathbf{i}\right\} + \left[1/\gamma M^2\right](\partial p/\partial z + \mathrm{Bo}\rho) = 0, \tag{3.21c}$$

$$p = \rho T \text{ and } \mathrm{St}\ \mathrm{D}(p/\rho^{\gamma})/\mathrm{Dt} = 0, \tag{3.21d, e}$$

where

$$\mathrm{St}\ \mathrm{D}/\mathrm{Dt} = \ \mathrm{St}\ \partial/\partial t + \mathbf{v}.\mathbf{D} + w\partial/\partial z, \text{ and } \mathbf{v} = u\mathbf{i} + v\mathbf{j};\ \mathbf{D} = \partial/\partial x\mathbf{i} + \partial/\partial y\mathbf{j}.$$

For the above system of equations (3.21a)–(3.21d, e), initial conditions must be written at $t = 0$, and we observe that the initial data for these equations *need not fit the hydrostatic balance (as a general rule!)*. In particular, the vertical velocity *need not be* $O(\varepsilon_0)$, with respect to the horizontal one, so that in order to consider the most general case, *it must be assumed* that, at the initial initial time, $\varepsilon_0$ w is of order O(1):

$$t = 0: \quad \mathbf{v} = \mathbf{V}^\circ, \; \varepsilon_0 w = W^\circ, \; p = P^\circ \text{ and } \rho = R^\circ, \tag{3.22}$$

where $\mathbf{V}^\circ$, $W^\circ$, $P^\circ$, and $R^\circ$ are given functions of z and also of the horizontal positions.

On the other hand, when considering the Kibel primitive hydrostatic equations (3.19b)–(3.20), *we must give*, in reality, *only* the initial values of $\mathbf{v} = (u, v)$ and $\rho$! Indeed, the initial value of $\rho$ yields, from the hydrostatic balance (3.19d), the initial value of the pressure p from $\partial p/\partial z = -Bo\rho$, while the relation (3.19e) yields the initial value of $T = p/\rho$. It seems *clear* that these initial values of **v** and $\rho$, *for the Kibel primitive hydrostatic equations* (3.19a)–(3.19f), *have nothing to do with the corresponding full initial conditions* (3.22) *for the full non-hydrostatic system of* Eulerian *equations* (3.21a)–(3.21d, e).

Consequently, we get as initial conditions for the Kibel primitive equations (3.19a–3.19f)

$$t = 0: \mathbf{v} = \mathbf{v}^\circ \text{ and } \rho = \rho^\circ, \tag{3.23}$$

where $\mathbf{v}^\circ$ is different from $\mathbf{V}^\circ$ and $\rho^\circ$ is different from $R^\circ$. Two of the initial conditions in (3.22) have been lost during the limiting process (3.20) and two questions arise (in the framework of a typical unsteady adjustment/transient local problem near the initial time):

1. How have these initial conditions been lost?
2. How are $\mathbf{v}^\circ$ and $\rho^\circ$, in (3.23), related to the initial data in (3.22) at $t = 0$?

Regarding the first question, the answer is obvious, and in fact, according to (3.19d) and (3.19e), in the above Kibel primitive system of equations, the pressure p is *related* to the density $\rho$ and T, while the vertical component of the velocity w—as noticed for the first time in the book <7> by *Richardson (1922)*—is *computed* by the process of solving the Kibel system of primitive equations.

All this holds true at the initial time as well! As a matter of fact, if we consider the Kibel primitive equations (3.19a)–(3.19f), and *rewrite* them *in the pressure coordinates developed by Eliassen* (1949) (and see <8>), they become:

$$\mathbf{D}.\mathbf{v} + \partial\varpi/\partial p = 0, \tag{3.24a}$$

$$\mathrm{St}\,\partial\mathbf{v}/\partial t + \mathbf{v}.\mathbf{D}\mathbf{v} + \varpi\partial\mathbf{v}/\partial p + [(1/Ro) + \beta y](\mathbf{k}\times\mathbf{v}) + (Bo/\gamma M^2)\mathbf{D}H = 0, \tag{3.24b}$$

$$T = -\,Bop\,\partial H/\partial p, \tag{3.24c}$$

$$\mathrm{St}\,\partial T/\partial t + \mathbf{v}.\mathbf{D}T + \varpi\{\partial T/\partial p - [(\gamma - 1)/\gamma](T/p)\} = 0. \tag{3.24d}$$

This is a *quasi-incompressible* form, where $\varpi = Dp/Dt$ is the the total derivative of the pressure, with *H* the local height above the *flat ground* of an *isobaric surface*, such that:

$$z = H(t,\ x,\ y,\ p) \text{ is a dependent variable, while,} \tag{3.24e}$$

$$St\ \partial/\partial t = St\ \partial/\partial t + \mathbf{v}.\mathbf{D} + \varpi \partial/\partial p. \tag{3.24f}$$

For the above *primitive à la Eliassen Eulerian Kibel equations* (3.24a)–(3.24d), with (3.24e, 3.24f), there are again *only two possible initial conditions at* $t = 0$, since the above system of primitive equations (3.24a)–(3.24d) is only of *second order in time t.*

For these primitive equations (3.24a)–(3.24d), we write also the following *slip condition* on the *flat ground*:

$$\text{on}\, H = 0:\ \varpi = Bo\ (p/T)\ [St\ \partial H/\partial t\ + \mathbf{v}.\ \mathbf{D}H]. \tag{3.25}$$

As a matter of fact, if we consider the above Kibel primitive system (3.24b–3.24e) *written à la Eliassen,* we observe that, the two main unknowns are, $\mathbf{v}$ and H.

This situation is *slightly reminiscent* of the one which occurs in classical *boundary layer theory*, if one considers that $\varpi$ is deduced from $\mathbf{v}$ by a kind of divergence-free condition, by way of (3.24a). We should keep in mind that $\varpi$ and H are known, apart from an integration constant, when T is known. Hence, w, the vertical component of the velocity, is not a primary unknown function, in the sense that it may be computed aftewards, through use of the above relation (3.24f) for $\varpi = Dp/Dt$, when H and $\varpi$ have themselves been computed from the knowledge of $\mathbf{v}$, T, and the pressure p.

### 3.4.4 *Unsteady Adjustment to Hydrostatic Balance, According to* **Guiraud** *and* **Zeytounian (1982)**

At the end of this rather lengthy Sect. 3.4, let us answer the second question above, and derive the *local (inner* in *time) working model equations, which are significant near the initial time*. This derivation will *illustrate* the importance of *matching* and the *role* of the local/inner limiting process near $t = 0$, in the context of the *local working model* and associated *adjustment problem, when we seek valid initial conditions* (3.23) *for the Kibel system of primitive equations (3.19a)–(3.19f)!*

In the present case, it is *fairly* obvious that the proper rescaling of the time and dependent functions is

$$t = \varepsilon\tau,\ \ z = \zeta,\ \mathbf{v}^{\circ} \equiv \mathbf{v}^{I},\ \ \varepsilon w = w^{I},\ \ p^{\circ} \equiv p^{I},\ \ \rho^{\circ} \equiv \rho^{I}, \tag{3.26a}$$

when we observe that the *horizontal position (x and y, which play only the role of parameters) do not play any role* in the local problem of *adjustment to hydrostatic balance* (3.24c). The associated local in time (inner) limiting process near $t = 0$ is

$$\varepsilon \to 0, \text{ with } \tau \text{ and } \zeta, \text{ and also } w^I \text{ fixed.} \tag{3.26b}$$

We take the set $\mathbf{v}_0{}^I$, $w_0{}^I$, $p_0{}^I$, $\rho_0{}^I$ for the limiting values, functions of $\tau$, $\zeta$, and the horizontal position as parameters. It is then straightforward (the *curious* reader will easily convince himself!) to derive the following set of *local initial layer, non-hydrostatic, leading-order, significant,* companion *equations* to (3.24a–3.24d) for $w_0{}^I$, $p_0{}^I$, $\rho_0{}^I$, from the full tangent, non-hydrostatic equations (3.21b)–(3.21d, e):

$$\gamma M^2 \rho_0{}^I (\partial w_0{}^I/\partial\tau + w_0{}^I \partial w_0{}^I/\partial\zeta) + \partial p_0{}^I/\partial\zeta + \rho_0{}^I = 0, \tag{3.27a}$$

$$\partial \rho_0{}^I/\partial\tau + \partial(\rho^I{}_0 w_0{}^I)/\partial\zeta = 0, \tag{3.27b}$$

$$(\partial/\partial\tau + w_0{}^I \partial/\partial\zeta)\,[p_0{}^I/\rho_0{}^I] = 0, \tag{3.27c}$$

with

$$(\partial \mathbf{v}_0{}^I/\partial\tau + w_0{}^I \partial \mathbf{v}_0{}^I/\partial\zeta) = 0, \tag{3.28}$$

when $St \equiv 1$, with $t_0 \equiv L_0/U_0$ and $Bo \equiv 1$ ($H = RT_{st}(0)/g$).

The above system of *three (adjustment) equations* (3.27a)–(3.27c) for $w_0{}^I$, $p_0{}^I$, and $\rho_0{}^I$ is identical to the system for *one-dimensional unsteady vertical gas dynamics in the atmosphere*. In fact, the system of local equations (3.27a–3.27c)–(3.29a) is the exact local form of the full system (3.21b)–(3.21e)!

Once $w_0{}^I$ has been obtained by solving the system (3.27a–3.27c) with the initial conditions (see 3.22)

$$\tau = 0: \; w_0{}^I = W^\circ, \; p_0{}^I = P^\circ \text{ and } \rho_0{}^I = R^\circ, \tag{3.29a}$$

and proper boundary conditions on the ground and at infinity (see 3.30a and 3.30b), we may use the transport (3.28) to compute $\mathbf{v}_0{}^I$, using the initial condition

$$\tau = 0: \; \mathbf{v}_0{}^I = \mathbf{V}^\circ, \tag{3.29b}$$

since equations (3.28) merely says that $\mathbf{v}_0{}^I$ is convected without change, vertically, with velocity $w_0{}^I$. The set of equations (3.27a–3.27c) was solved numerically by *Outrebon (1981)* (see pp. 94–101 in Chap. 5 of <3>, cited in Chap. 2), using the *slip condition*

$$w_0{}^I = 0 \text{ at } \zeta = 0, \tag{3.30a}$$

and enforcing with

$$\partial w_0{}^I/\partial\zeta = 0 \text{ and } \partial p_0{}^I/\partial\zeta + \rho_0{}^I = 0, \tag{3.30b}$$

at the *maximum altitude* (20 km) used in the computational grid.

Note that the above *theory of the adjustment to hydrostatic balance* was published in the journal *Tellus* by *Guiraud* and *Zeytounian (1982)*. The numerical code used by *Outrebon* (in the context of a *French university thesis (1981)*) was inspired by a code due to *Lerat and Peyret (1973)*, which was itself a generalisation of work by *Fromm (1969)*.

In Chap. 5 by Zeytounian <3>, *cited in Chap. 2* entitled "*Meteorological Fluid Dynamics, the (curious!) reader can find various results and conclusions from Outrebon's numerical computations. The final result may be stated very simply as follows: The horizontal component of the velocity and the specific entropy in the two sets of initial conditions are merely shifted vertically by the amount of vertical displacement during the whole process of the vertical, one-dimensional, unsteady adjustment to the balance equation!*"

## References

<1> Zeytounian RKh (1999) Well-posedness of problems in fluid mechanics. Russ Math Surv 54(3):479–564

<2> Zeytounian, Guiraud JP (1984) Asymptotic features of low Mach number flows in aerodynamics and in the atmosphere. In: Miller JJH (ed) Advances in computational methods for boundary and interior layers, 18–19 June 1984. Boole Press, BAIL III, Dublin

<3> Kibel IA (1957/1963) An introduction to the hydrodynamical methods of short period weather forecasting. Pergamon, New York (1963) translated from Russian edition, Moscow (1957)

<4> Zeytounian RKh (1974) Arch Mech Stosow 26(3):499–509

<5> Guiraud JP, Zeytounian RKh (1979) Geophys Astrophys Fluid Dyn 12 (1/2):51–72

<6> Guyon E, Hulin J-P, Petit L, Mitescu CD (2001) Physical hydrodynamics. Oxford University Press, Oxford

<7> Richardson LF (1922/1966) Weather prediction by numerical process. Cambridge (Reprinted by Dover Publications in 1966)

<8> Eliassen A (1949) In: Geofys Publ, vol 17, p 3

# Chapter 4
# Some Concluding Remarks About Part I

The present Part I, containing Chaps. 2 and 3, can be viewed as a *short introduction to some special analytical aspects of asymptotic fluid dynamics*. With Chap. 1 *devoted to the discovery of the NSF system of equations*, these three chapters, written in a more expository style, are essentially self-contained and make up a text which would require minimal amplification for a one-semester course for post-graduate students in applied mathematics and research engineers working in numerical computations and also in simulation of stiff fluid dynamics problems. Of course, this amplification can be realized using the *two companion books published in* 2012 and 2014 mentioned in preface.

In any case, thanks to Chap. 8, which constitutes Part III of the book, the reader will find in Sects. 8.1–8.9 a wide *range* of different *applications* of the *RAM approach* in *Newtonian fluid dynamics*, and it seems to me that this Part I with Part III will give the reader an *alternative and challenging view of fluid dynamics*. On the other hand, I do not think that it is a *suitable textbook* for a regular class, being more suited to individual researchers seeking entry, or as a tutorial by a teacher with some *PhD candidates* in the area, than as a conventional text for a lecture course.

Clearly, it is also suited to *self-study by an advanced student, one that is curious and highly motivated,* and it is important to *understand* that *asymptotic scaling techniques and the reduced, approximate reduced forms (working models) of the general equations in fluid dynamics (Euler, Navier, or NSF equations), that emerge from the application of the RAM approach provide a logical systematic method for obtaining the necessary simplified—numerically stiff—flow models, which in most cases still have to be solved with the help of super-high-speed computers!*

Concerning *scaling*, *dimensionless parameters,* and *reduction* of the full unsteady NSF equations to a system *without dimensions*, but with various *dimensionless reference parameters*, we observe that one of the most *fruitful uses* of scaling is the hierarchy of comprehensive and rich models it provides.

In most applications, such overcomplete models are not always truly adapted to the relevant multi-scale problem. The order of magnitude of the various



R.Kh. Zeytounian, *Challenges in Fluid Dynamics*,
DOI 10.1007/978-3-319-31619-2_4

contributions or effects can be estimated from a suitable inherent scaling—see, for example, the Chap. 7 (entitled *The Art of Modeling*) in the book by Mattheij, Rienstra, and ten Thije Boonkkamp: *Partial Differential Equations: Modeling, Analysis, Computation*, SIAM 2005, Philadelphia—leading to a hierarchy on which we can base asymptotic modeling.

Note that, in the above Sect. 3.4, a typical but not universal approach is worked out, since problem-dependent choices sometimes have to be made.

## 4.1 An Another Point of View Concerning the Dimensionless Equations

Below, we recall the mass, momentum, and energy conservation equations (a system of three equations), neglecting the radiation R in the energy equation. In particular, we consider the following simple system for ρ, **u**, and T, where gravity (Bo) and Coriolis (Ro) forces are not taken into account in the momentum equation (4.1b):

$$D\rho/Dt = -\rho\nabla . \mathbf{u}, \tag{4.1a}$$

$$D\mathbf{u}/Dt = -\nabla p + \nabla . \underline{\mathbf{T}}, \tag{4.1b}$$

$$\rho CpDT/Dt = Dp/Dt - \nabla . \mathbf{q} + \underline{\mathbf{T}} : \nabla \underline{\mathbf{u}}. \tag{4.1c}$$

Below a subscript "$_O$" refers to the primary variable.

The typical time is written as the reciprocal of a frequency f, but there are more candidates for the scaling parameter. Indeed, the typical frequency may be enforced by an external source or instability, or it may be inherited from the intrinsic hydrodynamic frequency $U_O/L_O = t_O$ or $k_O/\rho_O CpL_O{}^2$, which is the reciprocal of a diffusion time. On the other hand, we know that the typical sound speed squared $C_O{}^2$ scales with both $p_O/\rho_O$ and $\Delta p/\Delta\rho$ (not exactly, but to within an order of magnitude), where Δ here denotes a typical difference. We can thus take $p_O = \rho_O C_O{}^2$ and $\Delta\rho = \Delta p/C_O{}^2$. Note that $k_O$ and $\mu_O$ are both constants. Below, in equations (4.4a)–(4.4c), we have four new parameters:

$$Eu = \Delta p/\rho_O U_O{}^2, \tag{4.2a}$$

$$Fo = (k_O/f)\,/\rho_O CpL_O{}^2, \tag{4.2b}$$

$$He = f\,L_O/C_O, \tag{4.2c}$$

$$Ec = U_O{}^2/Cp\Delta T, \tag{4.2d}$$

which have the following physical interpretations:

The Euler number compares the available pressure difference with the typical pressure difference that can be expected from hydrodynamical effects alone.

The Fourier number compares a time scale 1/f with the typical time necessary for the diffusion of heat over a distance $L_O$.

The Helmholtz number compares the typical wavelength of sound with the size of a scattering object or a source, and tells us a lot about the effectiveness of the scatterer or source.

The Eckert number compares the kinetic energy of the flow with the available difference in enthalpy.

We observe also that the following relations are satisfied:

$$\mathrm{He} = \mathrm{M\ St}, \quad \text{and} \quad \mathrm{St\ Fo\ Pr\ Re} = 1. \tag{4.3}$$

Finally, the new scaled equations are now, in dimensionless variables:

$$\mathrm{Eu}\,\mathrm{M}^2(\mathrm{St}\ \partial\rho/\partial t + \mathbf{u}\,.\,\nabla\rho) = -\,\rho\nabla\,.\,\mathbf{u}, \tag{4.4a}$$

$$(\mathrm{St}\,\rho\ \partial\mathbf{u}/\partial t + \rho\mathbf{u}\,.\,\nabla\mathbf{u}) = -\mathrm{Eu}\nabla p + (1/\mathrm{Re})\nabla\,.\,\underline{\mathrm{T}}, \tag{4.4b}$$

$$\mathrm{St}\ \rho\ \partial T/\partial t + \rho\mathbf{u}\,.\nabla T = \mathrm{Eu\ Ec}\,(\mathrm{St}\partial p/\partial t + \mathbf{u}\,.\,\nabla p) + (1/\mathrm{P\acute{e}})\underline{\mathbf{T}}:\underline{\nabla\mathbf{u}}. \tag{4.4c}$$

When the numbers or parameters in the above equations become small or large, we may derive various reduced working models, as we did for the atmospheric motions in Sect. 3.4. In such a case we assume that the corresponding terms either vanish or dominate completely.

Other terms are sometimes best taken equal to one—in particular if one of the scales involved is inherent rather than externally enforced. But it seems to me that the derivation of rational consistent approximate working models is often more subtle! First, as already mentioned, the experienced reader will be aware of the fact that the NSF equations, even written in dimensionless form, are not sufficient for discussing fluid flow problems, and this seemingly innocuous remark has far-reaching consequences! As a matter of fact:

*Modeling is not a purely mathematical exercice!*

In Chaps. 5 and 6, and especially in Chap. 8 where this will be very well demonstrated via the derivation of numerous valid but approximate working models, the derivation of such *reduced models* must imperatively take into account the *associated dimensionless initial and boundary conditions, which to a large extent characterize the particular physical nature of the NSF problem!*

## 4.2 The Singular Nature of Fluid Dynamics

Unfortunately, most of the *relevant* engineering computations seems to be *based* on relatively ad-hoc models that are rife with internal inconsistencies. In particular, we have already been *confronted* with the "*disappearance*" of some *partial derivative terms in the derived models!* As a consequence, in various local regions (mainly *close to t = 0*, and in the *vicinity of the wall* of a body in motion), one must consider various *unsteady adjustment problems and boundary layers*!

This is precisely the main reason for the application of our RAM approach, which allows a careful inspection of the *singular nature* of the relevant *limiting processes,* with *time* and *space coordinates fixed,* and provides a way to derive the associated local working models, and via matching, to *relate the principal model with these local models*! Although this is usually a very difficult exercise, it also seems to me "unavoidable" if we want to obtain at least an approximate valid leading-order reduced working model in place of the very stiff full NSF problem.

Concerning the singularity near $t = 0$, in the framework of the RAM approach, the logical way to solve the associated local/short-time unsteady problem, with $\tau = t/\varepsilon$, $\varepsilon \ll 1$, is to consider an *initial layer* near $\tau = 0$, as formulated in Sect. 3.4 for the local problem of the *unsteady adjustment to hydrostatic balance*, where matching plays the main role:

$$\mathrm{Lim}_{\text{when } \tau \to \infty} \text{ (in short-time local adjustment model equations)}$$

$$= [\text{for principal evolution working model}]_{\text{at } t = 0}. \tag{4.5}$$

The above relation provides a way to bring together the short-time local adjustment model and the principal evolution working model, to build an approximate valid leading-order reduced working model in place of the original stiff full NSF problem. But it is also important to note that, depending on the kind of physical NSF problem, there are two main kinds of behavior for the $\mathrm{Lim}_{\text{when } \tau \to \infty}$. Either the system tends *towards a limiting steady state* (and in such a case *matching is ensured*!) or an *undamped set of oscillations appear*, and one must then *apply* a *multi-scale asymptotic approach* (see Sect. 8.9).

Important and very interesting problems arise when modelling the interaction between slow conduction-dominated thermal effects and much faster acoustic mechanical processes, and these can have significant effects which are often beyond suspicion. These problems require systematic examination of the complete unsteady NSF equations for a viscous and heat-conducting compressible gas. For instance, it seems that even for small linear disturbances generated in a gas by distributed sources of energy and/or forces, a small conduction-dominated region would form around an impulsive concentrated heat source.

In the paper by *Radwan* and *Kassoy (Physics of Fluids, 1984),* the reader will find a good introduction with various references concerning the thermoacoustic phenomena that occur in interior and exterior flows, and which frequently play a role in noise generation. In particular, in the above-mentioned paper, *the response of a confined gas to a thermal disturbance, with rapid boundary heating,* is carefully described in the spirit of the RAM approach. The main assumption is as follows: *The wall temperature is raised by an amount Q(t) of order O(1), with dQ/dt = O(1) and Q(0) = 0 at the boundary, and the model is based on asymptotic expansions relative to the limit ε → 0, where ε is the ratio of two times: acoustic time/conduction time!*

In relation to a *spectacular*, but *tragic event*, I would like to mention once again the thermal problem of *simulating the sudden ignition* that initiated the break-up and loss of the space shuttle Challenger during flight STS-51-L, which occurred on January 28, 1986. Clearly (at least for me), the *fluid dynamics simulation* of this disaster should be developed in *two stages*. The *first stage arises from* the *suddenness at* $t = t_{Ac}$ of the *appearance of flames during a very short period of time*, via the term $Q([t - t_{Ac}]/\varepsilon^{\alpha})$, with $\alpha > 0$, and see also the page 105 and (5.1a) and (5.1b). The second stage concerns the *thermal evolution process, over a longer period.* This second stage, in the *framework of high Reynolds numbers*, seems *well modelled* by a *regular classical Euler-Prandtl coupling* via the *boundary-layer evolution equations, but far from the initial time* where the data *for the NSF starting problem are given.*

In contrast, the initial, very short time stage is strongly related to a more complicated *singularity* in the wall temperature field *near* the "accident time" $t = t_{Ac}$, according to the RAM/deconstruction approach, which is carried out in Sect. 6.1, *assuming* $Re \gg 1$.

*For* a possible use of such a RAM approach via the local Rayleigh-Howarth viscous and heat conducting equations in a *small corner region I* (see Fig. 6.1), one must assume (with dimensionless time) that the *short time* near $t = t_{Ac}$ in the above temperature excess is such that

$$\alpha = 2, \text{ and } t - t_{Ac} \approx 1/Re, \text{ when it is assumed that } Re >> 1. \tag{4.6}$$

We also observe that these local *Rayleigh-Howarth* viscous and heat conducting equations, consistent in the corner region *near the initial time* and *close to the wall*, as a system of unsteady adjustment equations with the Prandtl boundary-layer evolution equations consistent far from the initial time $t = 0$, lead to *Euler-Prandtl-Rayleigh coupling*. These *three model coupling equations* are significant degeneracies of the full system of unsteady NSF equations when the Reynolds number is very large—for a vanishing viscosity—and can also be *useful*, not only in *thermoacoustics,* but also in *environmental fluid dynamics* (for, instance, for *predicting a forest fire front*).

*As a matter of fact, according to Part I, if we consider the full unsteady NSF system of equations with initial and boundary conditions as the fundamental mathematical basis for the investigation and numerical computation of Newtonian fluid dynamics problems, then we must clearly consider both the Euler (non-viscous) and Navier (incompressible) cases as two (effectively) important, but only approximate, reduced, significant working models, which must be derived from NSF when* $Re \to \infty$ *(vanishing viscosity) and* $M \to 0$ *(weakly compressible), respectively.*

Moreover, if we take into account the similarity relation (3.8a), with two specific situations (3.8c) and (3.8d), then we open up a very broad road towards a consistent family of *second-order working models*, companion to well known useful models, considered in the framework of our RAM approach as leading-order models! But for this, one must first formulate the problem to hand as an NSF system, where the Reynolds and Mach numbers are relevant! Such a case is considered in Sect. 8.5.

## 4.3 Unsteady Fluid Flows

On the other hand, one should bear *in mind* that the *unsteady and steady fluid flow cases must be considered separately*, and this observation is not trivial. For instance, in the framework of the Euler system of *unsteady* compressible equations:

$$\begin{aligned} D\rho/Dt + \rho\nabla\,.\,\mathbf{u} &= 0, \\ \rho D\mathbf{u}/Dt &= \rho f - \nabla p, \\ DS/Dt &= 0, \end{aligned} \tag{4.7}$$

by definition ($D/Dt = \partial/\partial t + \mathbf{u}.\nabla$), the *specific entropy* $S$(t, **x**) in 3D Eulerian baroclinic fluid flow is a *Lagrangian invariant* ($DS/Dt = 0$), and *S is conservative along the trajectories*.

On pp. 83–86 of *Zeytounian* [13], cited in "Introduction", the reader will find various results concerning these *Lagrangian invariants*—in particular, w*e have a system of five Lagragian invariants* associated with the above Euler (4.7) unsteady system of equations. In particular, the first one is $L_1$ (t, **x**) such that

$$DL_1/Dt \equiv D/Dt[(1/\rho)\Omega\,.\,\nabla S] = 0, \tag{4.8a}$$

where $\boldsymbol{\Omega} = \nabla \wedge \mathbf{v}$, and for the velocity **u** we write

$$\mathbf{u} = \nabla H + \mathbf{v}, \text{ with } dH/dt = (½)|\mathbf{u}|^2 - U - h. \tag{4.8b}$$

In fact, the following functional relation is true:

$$(1/\rho)\mathbf{v}\,.\,(\nabla S \wedge Ln) = Ln + 1, \text{ with } n = 1, 2, 3, \ldots. \tag{4.9}$$

However, among the infinitely many Lagrangian invariants (4.9), only two are *linearly independent*. In particular, if $\boldsymbol{\Omega} \equiv \boldsymbol{\omega}$, with $\omega = \nabla \wedge \mathbf{u}$, we obtain the well-known *Ertel potential vorticity* from $L_1$.

In Zeytounian <1>, the reader can find an application to the *calculation of 3D rotational Eulerian fluid flow* [expounded, in 1978, in *Tbilisi* (Soviet Georgia), *VI International Conference on Numerical Methods in Fluid Dynamics*].

But the *main cause* of such a *distinction* between *unsteady and steady fluid flows* relates to the constraints due to initial and boundary conditions. In particular, in the two limiting cases mentioned above, viz., $Re \to \infty$ and $M \to 0$, one must specify whether the time and space coordinates are fixed or not. For instance, when $Re \to \infty$, in the framework of a full unsteady NSF initial-boundary value aerodynamic external problem, if the time and space coordinates are fixed, we first have the classical Euler system of unsteady non-viscous equations—but this Euler reduced system is only a valid approximate system far from the boundary where the conditions for the velocity vector **u** and temperature T are prescribed in the framework of an initial-boundary value NSF problem!

In order to obtain, at least to leading order, a significant reduced approximate and useful working model for the numerician, one must take into account a second $Re \to \infty$, but with the new, vertical-to-boundary space coordinate $\zeta = z/\varepsilon$ fixed ($\varepsilon^2 = 1/Re$)—significant in the vicinity of the boundary – and the resulting reduced approximate model is the famous *Prandtl boundary layer system of unsteady equations*! The coupling between Euler and Prandtl is realized via the simplified matching condition, to leading order, by application of the relation (3.9c) in Sect. 3.3. But in the unsteady case, as shown below in Sect. 6.1, near the initial time, where the initial conditions are imposed, the *Prandtl boundary layer system of equation is in reality singular! As a consequence, the usual regular Euler-Prandtl coupling is consistent/valid only in the case of a steady fluid flow problem.*

In particular, near the initial time, one must consider consider a third limit, $Re \to \infty$, but with the new vertical-to-boundary space coordinate $\zeta = z/\varepsilon$ fixed, *and also* a *new short time* $\tau = t/\varepsilon$ fixed ($\varepsilon^2 = 1/Re$). The resulting reduced approximate model obeys the unsteady one-dimensional inviscid gas dynamics equations, where the horizontal coordinates x and y play the role of parameters! Unfortunately, these unsteady one-dimensional inviscid gas dynamics equations are only a *degenerate form* of the so-called *Rayleigh* unsteady-state one-dimensional, viscous, compressible, and heat-conducting equations, *close to the initial time and in the vicinity of the boundary*, in a *small corner region* ($\zeta^* = \zeta/\varepsilon$, and $\tau^* = \tau/\varepsilon$) of time-space!

A more careful inspection shows that the *matching between this corner region and the evolution in time of the boundary layer region* (far from $t = 0$), can be realized through an *intermediate fifth region*, again derived from the full unsteady-state NSF starting equations. These equations for this fifth region are derived, *on the one hand*, when we carry out the classical boundary layer approximations on the Rayleigh equations, and *on the other hand*, when we carry out the approximations used to derive the Rayleigh equations on the Prandtl boundary layer equations!

Concerning the *limiting process* $M \to 0$, considered in Sect. 6.2 (see Fig. 6.2), it mainly concerns the derivation of *Navier-Fourier* limit equations, and the equations of *acoustics*!

As already mentioned, on pp. 78–88 of our 2004 book [14], cited in "Introduction", the reader can find (see p. 79) a dimensionless formulation of the *compressible Rayleigh problem* by the *one-dimensional unsteady-state NSF equations* for the velocity components u and w, and thermodynamic functions p, ρ, T, dependent on t and z, with $St \equiv 1$ and also $Bo \equiv 0$.

In this system of Rayleigh equations, from the continuity equation,

$$\partial\rho/\partial t + \partial(\rho\, w)/\partial z = 0, \Rightarrow \rho = \partial\phi/\partial z \text{ and } w = -(1/\rho)\partial\phi/\partial t, \quad (4.10a)$$

where the function $\phi(t, z)$ is a "*particle function*", one can rewrite the Rayleigh problem in terms of the new coordinate and new time ($\phi$, $\theta$):

$$\phi = \int_0^z \rho\, dz', \phi = 0 \text{ at } z' \equiv z = 0, \quad (4.10b)$$

normal to the flat plate, and $\theta \equiv t$. We thereby obtain the following set of equations:

$$\begin{aligned}
&w = \partial z/\partial\theta,\ z = \textstyle\int_0^{\phi}(1/\rho)d\phi',\ \rho = p/T,\\
&\partial u/\partial\theta = (1/Re)\partial(p(\partial u/\partial\phi))/\partial\phi,\\
&\partial p/\partial\phi = \gamma M^2\{(4/3Re)\partial(p(\partial w/\partial\phi))/\partial\phi - \partial w/\partial\theta\},\\
&\partial T/\partial\theta = (1/Pr\,Re)\partial(p(\partial T/\partial\phi))/\partial\phi + [(\gamma-1)/\gamma]\ \partial p/\partial\theta\\
&\qquad + M^2[(\gamma-1)/Re]p\left[(\partial u/\partial\phi)^2 + (4/3)(\partial w/\partial\phi)^2\right],
\end{aligned} \tag{4.10c}$$

while the initial and boundary conditions are

$$\theta = 0 \text{ and } \phi \to +\infty - : u = w = 0,\ p = T = 1, \tag{4.11a}$$

$$\theta > 0 \text{ and } \phi = 0 : u = 1, w = 0,\ = T = 1 + \tau^{\circ}, \tag{4.11b}$$

because, when $z \to +\infty$, $\rho \to 1$, and consequently $\phi \equiv z$ at the limit, while $z \to +\infty$ is equivalent to $\phi \to +\infty$. The above initial-boundary value problem (4.7c) with conditions (4.8a, b) is investigated when

$$Pr = 1,\ Re = O(1),\ \text{and } M^2 \to 0, \tau^{\circ} \to 0, \tag{4.11c}$$

and the similarity relation

$$\tau^{\circ} = \Lambda^{\circ} M^2,\ \text{where } \Lambda^{\circ} = \gamma R \Delta T^{\circ}/U^{\circ} = O(1) \tag{4.11d}$$

is satisfied. *We first consider* the case for a *small* $M^2$ *solution close to* the flat plate, but *far* from the initial time, which appears only as an *inner* asymptotic solution, and then the case for a *small* $M^2$ *solution, far from* a flat plate (an *outer* asymptotic solution), with t fixed and $z^* = Mz$! But *to match the inner one* with *the outer asymptotic* expansions for the pressure, we must include a term for p in the inner expansion that is proportional to $M^3$!

Clearly, the procedure only yields an asymptotic solution for the time $t = O(1)$, and is not valid near $t = 0$, because the solution derived in the last two cases would imply that *an infinite implulse per unit area is required to set the flat plate in motion!*

The main problem lies in *matching* a solution that is valid i*n the initial transient layer* with an *asymptotic solution that is valid for fixed time* $t > 0$, far from $t = 0$! In the paper by Howard <2>, the reader will find some pertinent discussion (in particular, the initial stage of motion is investigated in his Sect. 4).

Close to the initial time $t \leq 0$, one must first introduce a short time adapted to the initial transient layer, namely, $\tau = t/M^{\beta}$, with $\beta < 0$. With this short time $\tau$, the following initial limiting process is considered:

$$M \to 0 \text{ with } \tau \text{ and z fixed and } Re = O(1). \tag{4.12a}$$

Then with the associated asymptotic expansions

$$w = M^{\alpha} w_i + \ldots, \; (p, \rho, T) = 1 + M(p_i, \rho_i, T_i) + \; \ldots, \tag{4.12b}$$

and with $p = \rho\, T$, we recover the classical acoustic inviscid equations for $w_i$ and ($p_i$, $\rho_i$, $T_i$), if $\beta = 1$ and $\alpha = 0$, such that *close to the initial time we have a term proportional to M in the initial asymptotic expansion*!

## 4.4 Some Arguments Regarding the Boussinesq Equations

Finally, concerning the famous *Boussinesq approximation, we observe that*, in Vol. II of his 1903 monograph entitled "*Théorie Analytique de la Chaleur*", Joseph Valentin Boussinesq wrote:

> The density variations can be ignored except where they are multiplied by the acceleration of gravity in the equation of motion for the vertical component of the velocity vector.

A spectacular consequence of this Boussinesq observation (simply called the *Boussinesq approximation* by *Rayleigh* in 1916) is the possibility of working with a *quasi-incompressible* system of coupled dynamic (Navier) and thermal (Fourier) equations, with buoyancy as the main driving force.

The paper by Zeytounian <8>, cited in Chap. 2—*100 years later*—briefly discusses the *life of Boussinesq* and his *observation*, together with the applicability and validity of this approximation for various thermal, geophysical, astrophysical, and magnetohydrodynamic problems in the framework of a *Boussinesquian fluid dynamics*. On pp. 51–55 of the *companion 2014* book, the reader can find *my adventure when I obtained the first justification of this Boussinesq approximation and derived the associated Boussinesq equations for a reduced isochoric steady fluid flow*:

$$\begin{aligned} &\nabla \,.\, \mathbf{u} = 0, \\ &\rho \mathbf{u} \,.\, \nabla \mathbf{u} = \rho \mathbf{g} - \nabla p, \\ &\mathbf{u} \,.\, \nabla \rho = 0, \end{aligned} \tag{4.13}$$

when the *density* $\rho$ is *conservative* along the *streamlines* and we use the *two first integrals* (1.24a and 1.24b), in the simple 2D case (x, z), derived in Sect. 1.2.1 above, relating to the *lee-wave problem, over and downstream of a mountain.*

On pp. 160–170 of our book, 2002 [13], cited in "Introduction", the reader will find several figures showing the *configuration of streamlines above various mountains and under various weather conditions* (which are taken into account, in equation (4.14a), via the constant $K^2{}_0$), and below via the so-called *Long, 1953* model, in <3>, for the function $\delta$(x, z), in the steady 2D case:

$$\partial^2\delta/\partial x^2 + \partial^2\delta/\partial z^2 + K^2{}_0\delta = 0, \tag{4.14a}$$

with the following conditions for the case of *lee waves* in a bounded duct at $z = 1$, with a curvilinear bottom $z = \nu\, h(x)$:

$$z = \nu\, h(x) : \delta = \nu\, h(x), z = 1 : \delta = 0, \tag{4.14b, c}$$

$$x \to -\infty : \delta \to 0 \text{ and } x \to +\infty : |\delta| < \infty. \tag{4.14d, e}$$

In the numerical computation of the above (nonlinear) problem (4.13)–(4.14a–e), the most serious difficulty arises from the *practical impossibility* of achieving such horizontal (relative to x) conditions at ± infinity!

The above problem must be treated as either a *Cauchy problem* or a *boundary-value problem* within a *limited interval*—but in both cases, this process entails the regrettable consequence of bringing in some of the eigenfunctions pertaining to the above *Helmholtz* equation (4.11a) as a parasite in the solutions. In *Journal Physics of Fluids* (Supplement II, *1969*, II-46-II-50), I suggested a method for filtering these parasitic solutions. Some results of numerical computations are shown in Fig. 4.1a and Fig. 4.1b from p. 167 of [13], cited in "Introduction".

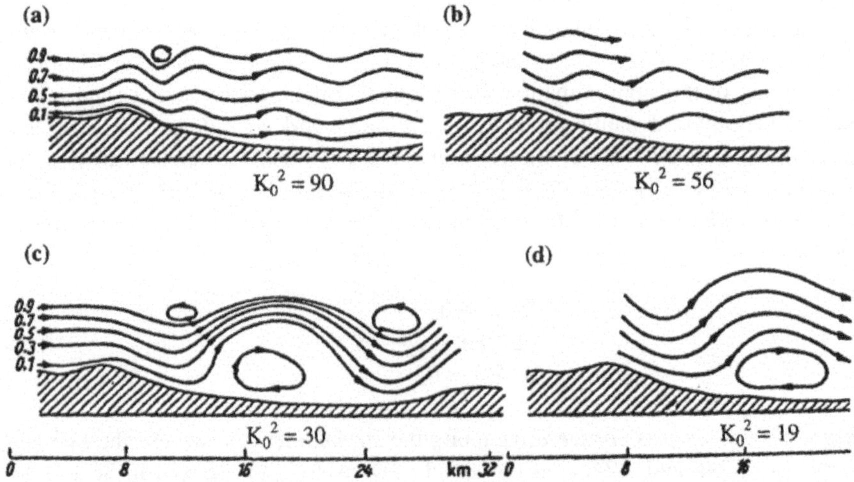


**Fig. 4.1a** Relief waves above Sierra Nevada for various values of $k_0^2$

(e)

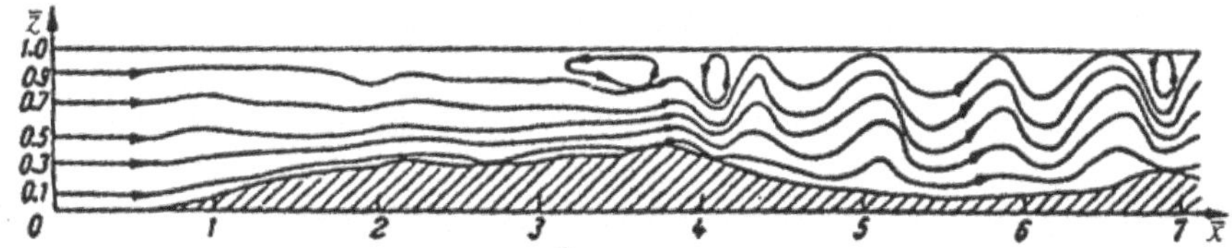

$K_0^2 = 60$

**Fig. 4.1b** Relief waves above Sierra Nevada for the value of $k_0^2 = 60$ a very exceptional case of the waves

## 4.5 The Ubiquitous Small Parameters

Obviously, the RAM approach requires one or more parameters to appear in the mathematical problem arising from the physical phenomenon under consideration. In [10], cited in "Introduction", 1986, devoted to *the general features of asymptotic modelling of flows*, the reader will find (pp. 10–14) some considerations concerning this "ubiquity"! Here, we give only a few typical examples.

The simplest situation is, of course, the appearance in the dimensionless *NSF problem* of the parameters mentioned in (3.2), and it concerns mainly the case of $M \ll 1$ and $1/Re \ll 1$. A more interesting case illustrates the occurrence of the small parameter in the boundary conditions, and it concerns *high aspect ratio wings, where the small parameter* is then the *reciprocal of the aspect ratio,* and it appears in the model when one goes into the details of writing the no-slip condition on the wing.

In various situations, the small parameter is built into a particular solution one is looking for, without being directly apparent in the formulation of the problem. A simple example concerns small amplitude waves of any kind. To be specific, for surface waves the small parameter is a ratio of lengths, namely the amplitude over the wavelength. A small parameter often arises in the relation beween two models, as in the theory of fluid-fluid interfaces with material properties!

It is interesting to observe that, whenever a researcher choses a numerical method for some situation, he is more or less confronted with a problem of asymptotic modelling!

## 4.6 A Guide for Classification

The asymptotic modelling/RAM approach plays a very important role in many aspects of fluid dynamics, but it would be nonsense to claim that modelling is mainly asymptotic! The importance of the *RAM approach* comes in at *three levels.*

*The level of general models*

Our thesis is that, very often, chapters in fluid mechanics may be organised through some model which is best got by asymptotic modelling. In particular:

I*nviscid* (*non-viscous*), *high Reynolds number* flows, which are often considered as models. From the outset, these need to be embedded in the more general model of *slightly viscous* (laminar) flows or flows with *slight friction* (turbulent flows), to which the asymptotic/RAM approach is applied.

*Creeping flows*, with numerous applications to *thin films* (lubrication, microdynamics, etc.) should be considered as flows *at low Reynolds number*. Entire books are devoted to creeping flows, the role of the Reynolds number being *ignored*, except in the first few expository pages.

*Incompressible* flows are seldom considered as flows *at small Mach number*. This may lead to something rather nonsensical, as when one deals with *incompressible aerodynamics*! Such phenomena, like sound produced by rather low speed flow, cannot be understood otherwise than as *low Mach number aerodynamics*.

*Rapidly rotating flows* which dominate applications of both industrial and geophysical nature, are indeed asymptotic models of flow at *low Rossby number*.

*Large scale models of flows*, of current use in *simulation for meteorological or oceanographic applications*, are extracted from asymptotic modelling, which explains the role of "*hydrostatic balance*".

A number of models for *flow in porous media* or flow *with suspensions* or *turbulent* flow should be considered as models obtained through some kind of *homogenization*.

*The level of local models*

A number of models used for the understanding of flows arise from asymptotic modelling applied to elucidate *behaviour* in some *localized region*. The examples are numerous and widely scattered across all the fields of fluid mechanics. We list a few of these—but without references (!).

In *wing theory*, the flow in the *vicinity* of either the l*eading*, the *trailing*, the *side*, or the *rounded edge* of the *planform* may be considered as *understandable* only on the *grounds of asymptotic modelling*!

*Laminar separation*, a challenge to fluid dynamics for seven decades, has been understood only as an application of the *triple deck asymptotic model*.

*Interfaces* endowed with material properties, or simply too thick to be considered as *pure discontinuities*, are best understood when considered as *thin layers embedded* within a *large scale flow*.

*Tightly wound rolled vortex sheets* may be grasped, independently of the question of their persistence against disintegration, by an application of *multiple scaling*.

The so-called *Riemann problem in gas dynamics*, especially when used in a numerical scheme like *Godunov's*, is actually an *asymptotic mode*l for small increases in time, after a definite time, where some information is given, including more or less *a departure from smoothness*.

Flows *over particles in suspension* within the flow, and their embedding in a large scale flow, are to be considered as l*ocal asymptotic models*.

*Cores* of *vortex filaments* and derivation of *algorithms for the motion* of the filaments are *asymptotic models.*

Finally, *entrance flows and subgrid scale modelling of turbulence are asymptotic models, the latter being an open field of research.*

*The level of global specific models*

It happens sometimes that the flow under consideration may as a whole be grasped via asymptotic modelling, using the RAM approach.

*Slender bodies* and the flows over them correspond to asymptotic models which have been widely used in *aerodynamics* or in *ship hydrodynamics.*

*High aspect wings* or *propellers* give useful applications of the concept of asymptotic modelling, capable of grasping the *whole of a flow.*

The flow within a *rotating engine* may be reduced to gluing together, via some asymptotic machinery, two apparently very different models, namely *cascade flow* and *through flow theories.*

Even the so-called *actuator disc model* may be considered *as part of the whole.*

*Thin film coating of a wire* by a very viscous liquid *may be tackled* as a whole through *asymptotic modelling*!

At the end of our 1986 paper with Guiraud in [10], cited in "Introduction", the reader can find a list of references organized by research theme:

*Vortex Flows in Rotating Machines,*
*Rolled Vortex Sheets,*
*Hydrodynamic Stability,*
*Atmospheric Flows,*
*Flow at Low Mach Numbers,*
*Miscellanea.*

There are also a number of references to either separate or joint authors devoted, with varying degrees of success, to the application of theses ideas in various fields of fluid dynamics, all being motivated during the years 1970–1986 by the need to solve, or understand a basis for the solution of, some technological problems involving fluid flow.

## References

<1> Zeytounian RKh (1979) Application of Lagrangian invariants to the calculation of three-dimensional rotational flows of a perfect fluid. LNP 90, Springer, Heidelberg, pp 594–598
<2> Howard L (1951) Some aspects of Rayleigh's problem for a compressible fluid. Q J Mech Appl 4(2):157–165
<3> Long RR (1953) Tellus 5:42–57

# Part II
# Deconstruction/RAM/Simulation Working Models and Their Classification

# Chapter 5
# Models Emerging from the Full NSF Problem via the RAM Approach

It has long been known that varied mathematical models consisting of partial differential equations (PDE), associated with boundary conditions (BC) and initial data (ID), can be used to solve a number of fluid flow problems. Let us mention a few of them: *thin air foil theory; hyposonic low Mach number fluid flows; subsonic, supersonic, and hypersonic fluid flows; boundary layers; low Reynolds number fluid flows; surfaces waves, acoustic waves, thermal convection....!*

*It is now always recognized that all these problems can be derived from the NSF system of PDEs associated with specific BC and ID, but since the direct numerical simulation/computation of a full NSF problem with BC and ID, is too time-consuming, it is convenient to appeal to the previously specified problems as models, each being appropriate for part of the full space-time domain, where the corresponding full NSF problem is formulated! The main advantage is to save time in numerical computation because of the "stiffness" of these fluid dynamics problems!*

*Let us call a system of PDEs with BC and ID a model—the main goal is modelling, namely replacing the full unsteady NSF (starting) model by various simplified reduced models corresponding to various sub-domains of the real fluid flow space-time domain, so that the appropriate combination (*via *matching) of their solutions gives at least a leading-order rational/significant approximation to the full solution* via *a numerical computation. But we must do this with a rational theoretical method, itself based upon a logical foundation—our RAM approach is just such a method. We may describe it by stating that it rests on two basic concepts: the first is "structuralism" and the second "deconstruction".*


R.Kh. Zeytounian, *Challenges in Fluid Dynamics*,
DOI 10.1007/978-3-319-31619-2_5

## 5.1 Structuralism and Deconstruction

*Deconstruction and structuralism are two concepts which arose in the schools of twentieth century Western criticism, and their* relationship is still an issue that needs to be systematically clarified, at least insofar as they are relevant here for our RAM approach! *Deconstruction* stems from *structuralism,* but deconstruction differs from structuralism in certain key features, being based upon a certain characteristic critique of structuralism! If *structuralism* were the father, then *deconstruction* might be regarded as the son.

Our main interest in structuralism concerns the "*infrastructure analysis*" of the NSF system of equations governing Newtonian compressible, viscous, and heat conducting fluid flows. This is an inquiry into deep structures rather than surface phenomena, and our activity consists of two main parts: dissection, in which we cut the starting NSF system of equations *into several parts,* and the search for certain partial fragments whose causes have a deeper meaning.

Of course, a fragment has no meaning by itself, but the slightest variation would change the final meaning of the whole NSF system, as a central system in Newtonian fluid dynamics, which gives the possibility of obtaining information about the origins and classification of a useful family of models existing in traditional fluid dynamics!

Deconstruction, i*nitiated by the French philosopher and critic Jacques Derrida,* is for me a particular method of analysis and argument involving the more detailed and deeper meaning of the NSF equations! In this sense, deconstruction is simply a way to get a deeper understanding of the more puzzling aspects of those equations!

The first states that, *according to Zeytounian's idea (!),* the various terms occurring in the mathematical definition of the NSF problem *have meaning only when they are considered in relation to each other*. This assumes that the problem *does not generate chaos* and *excludes turbulence*. To be more specific, any derivative occurring in the full unsteady NSF system of equations, when considered varying within some limited space-time domain, should not take on numerical values ranging outside an interval whose amplitude exceeds a certain magnitude.

We recognize the usual numerical magnitude (NM), but it depends on the choice of units, so it is essential to formulate the given full unsteady NSF initial-boundary value problem in dimensionless form and to use various limiting processes relative to significant large or small non-dimensional parameters. And certainly, it is also necessary to associate suitable asymptotic expansions with the relevant limiting processes! This means that the RAM approach can be used to obtain approximate *reduced working models* for some very interesting and technologically and geophysically important flows in the atmosphere and oceans.

*Unfortunately, these flows are not easy to understand, and in addition these limiting processes in the framework of the NSF equations often turn out to be of non-uniform validity, and they break down in certain regions of the flow, usually near the initial time, where the initial date are given, and in the vicinity of the boundary of the wall of a body in motion.*

In some cases, also near the accident time $t_{Ac}$, at *all events when* a "*temperature disaster*" *arises*, the RAM approach must be used to construct physically meaningful local working models via the local limiting processes, introducing a short time t and a reduced vertical coordinate z in place of t and fixed x, y, z! On the other hand, one must take into account the fact that, while some conditions must be assumed for the behaviour at infinity for external fluid flows, this is not the case for internal fluid flows!

However, in the framework of *low Mach number asymptotics*, i*nternal unsteady compressible aerodynamics* has become a field of *intensive study*, for instance, when a gas is enclosed in a container W(t), yielding a bounded closed volume with an impermeable but possibly *deformable* (in *time*) *wall*, so that the motion in W (t) can be a function of time, and this *leads to a very complicated problem*. More specifically, we observe, the following. On the one hand, a *peculiar derivative* (of any order in the NSF equations) is not significant by itself from the point of view of the "*structuralism*", but only when it is considered *in relation with others*. On the other hand, *the terms* in the NSF equations are *divided into several groups* so that *one may associate a dimensionless common factor* to each, i.e., a *non-dimensional reference number*.

Indeed, *each non-dimensional number is strongly linked with a particular significant physical property in the dimensionless NSF initial-boundary value fluid flow problem*, and with the help *of the RAM approach*, a *simplified/reduced working model problem may be extracted from the full dimensionless unsteady NSF initial-boundary value problem by:*

"*eliminating any group with a dimensionless common factor which is estimated to be very small relative to unity.*"

Obviously, the above-mentioned *elimination/estimation* is often a difficult *task*, and it is established mainly from the *essential physical nature of the particular fluid flow under investigation. In fact, the modeling is typically a mathematical-physical RAM approach, and pure mathematics does not play any role*! As a matter of fact, during my own application of the RAM approach to the *famous Bénard problem* (considered in Sect. 6.4, but see also Chap. 1 of my 2009 book [25], cited in "Introduction"), I often attentively read various parts of the book *Physical Hydrodynamics*, by Guyon, Hulin, Petit, and Mitescu <6>, cited in Chap. 3, which is strongly inspired by the Lectures of *Pierre-Gilles de Gennes*.

Another fascinating engineering application of the RAM approach to a complicated *turbomachinery axial fluid flow problem*, with *the number of turbine blades per rotor* as a basic large parameter, is our asymptotic theory imagined in 1971 with J-P Guiraud at the ONERA (Chatillon, France) [19, 20], cited in "Introduction".

Unfortunately, we must stress once again that *one should not expect such an elimination process to be valid for the whole space-time domain covered by the original NSF problem*! We are thus confronted with a *process of deconstruction of the starting NSF system of initial and boundary conditions*, into a different *simplified/reduced problems associated* with various *space-time subdomains*.

Hence, the *real meaning* of an *NSF problem* depends on its *dimensionless forms with various non-dimensional reference parameters* (or numbers), and as mentioned in Sect. 3.2, depends on the nature of the problem and also on the region of space-time where one carries out the modeling!

The main *starting argument* in the *RAM approach* is *to detect*, in the initial *NSF fluid flow problem, at least one significant small* (or *large*) non-dimensional parameter. This is exemplified by simplified models with *vanishing* viscosity, *weakly* expansible liquids, *low* or *slightly* compressible fluids, etc., which may be improved using an *asymptotic expansion* or at least a *limiting process* to produce a *leading-order working model.*

A difficult *localisation problem* concerns the singular nature of these asymptotic expansions or limiting processes in *some space-time domains*, and "deconstruction" is once again advocated. The solution of the original problem is then obtained by *assembling the various approximate reduced working models* corresponding to different space-time regions of the relevant starting domain, where the initial-boundary value NSF problem is mathematically formulated.

A further difficult problem concerns recovery of these *local reduced working models* themselves *together* with the *main model* by the process of *matching*, using the MAE. Unfortunately, in the case of internal problems, the matching does not work, as is clearly illustrated by "*A low-Mach number flow affected by acoustic effects in a confined gas over a long time*" (see for example Chap. 5 in our book [24], cited in "Introduction"). In such a case, one must use a *multiple-time scale technique to investigate the long-time evolution of the rapid acoustic oscillations*!

It is obvious that, via *deconstruction*—a word invented by *Jacques Derrida*—our *RAM approach* presents us *unexpectedly* with the *possibility of understanding the intrinsic structure of the full NSF equations, and reveals the presence of a profound unity in the puzzle constituted by the numerous partial models in the framework of Newtonian classical fluid dynamics.*

To "*deconstruct*" the *full NSF system of equations,* we have the possibility (see Sect. 5.2) to join together these many partial models, used in classical Newtonian fluid dynamics! From such a process we *re-establish, in a well-ordered* and *unified family, the various partial Newtonian fluid flow working models.* It seems me that "pure mathematicians" must be ready to take up *some of the fruitful ideas* from such a *new vision of the NSF equations* for their rigorous abstract investigations relating to the existence and uniqueness of the solutions of certain reduced working models!

## 5.2 Origins and Classification of a Family of Working Models Emerging from the NSF Problem. Figures 5.1, 5.2, 5.3, 5.4 and 5.5

In our books written during the years 1974–2014, a large family of fluid flow problems was investigated via the RAM approach, and many of these working problems are summarized below with the help of two illustrat*ive sketches in* Figs. 5.1 *and* 5.2. *It is clear that "implicitly", at least for me, the full NSF system*

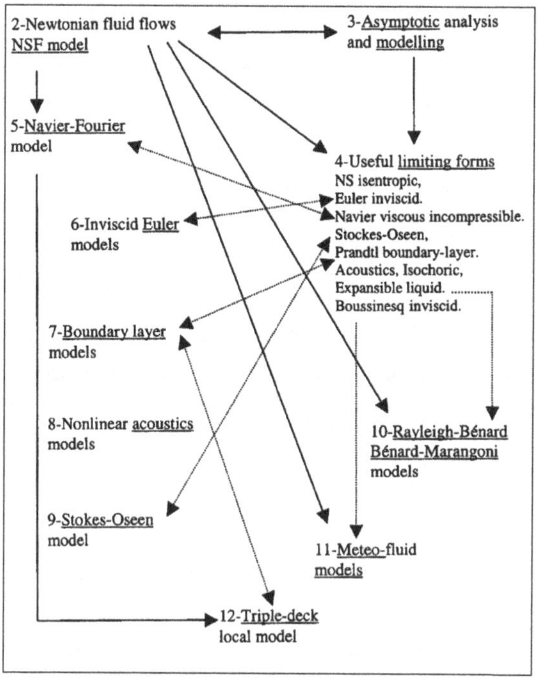


**Fig. 5.1** A large family of working models emerging from NSF

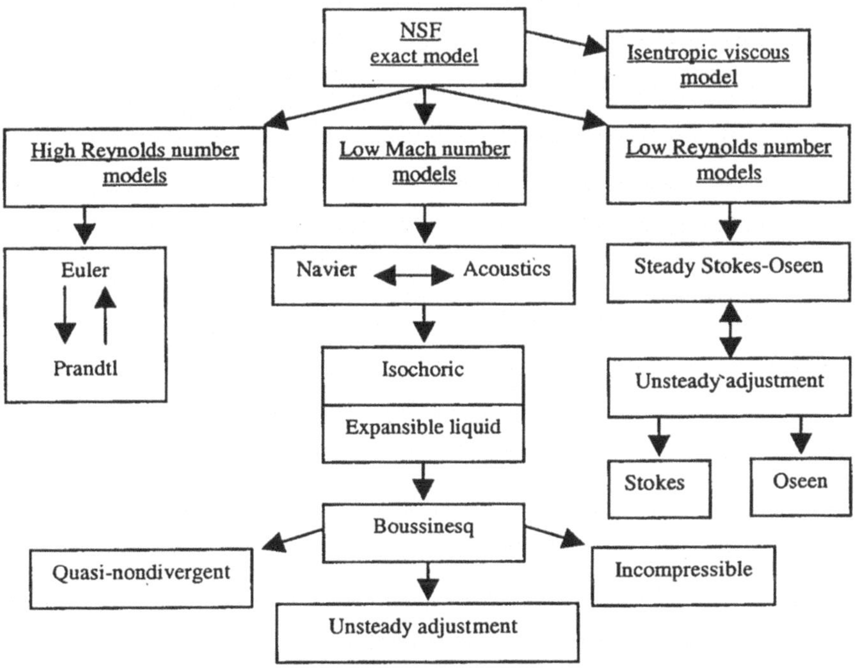


**Fig. 5.2** Various limiting forms of the NSF equation

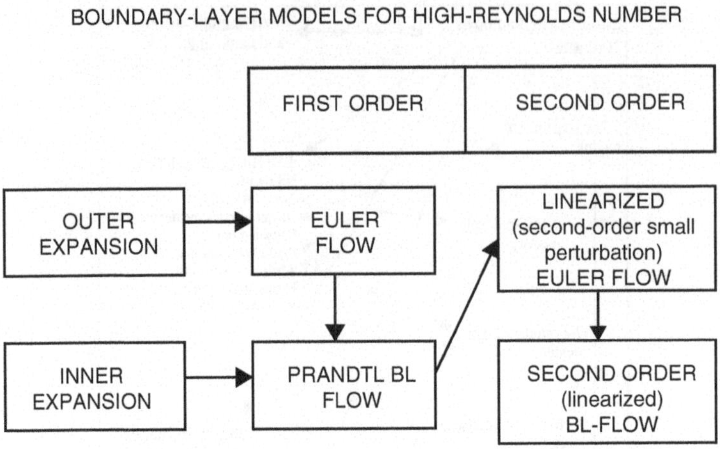


**Fig. 5.3** Euler–Prandtl non-interacting coupling

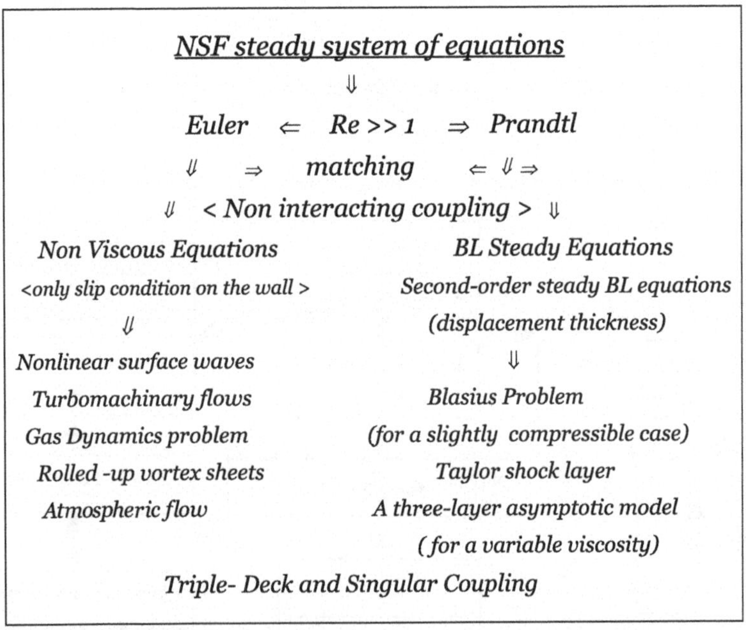


**Fig. 5.4** Various models with Re $\gg$ 1 asymptotics

*of equations is really a puzzle made up of very particular approximate reduced working model equations!* On the other hand, *I think that the most important problem* is actually the *possibility of getting a good understanding of the role of the full unsteady NSF equations* and the *possibility of practical applications of this system in numerical/computational fluid dynamics!*

Paradoxically, at the present time, *we are far from having a good idea of the close relation between this full NSF system of equations and the various useful*

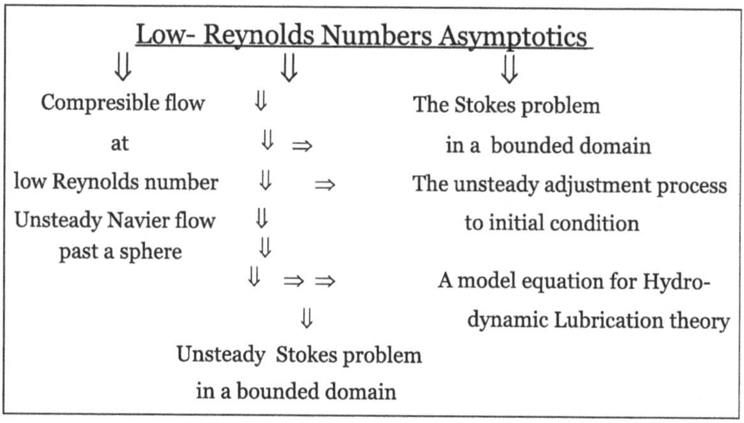


**Fig. 5.5** Re $\ll 1$ asymptotics

*model equations which are the main subject of university textbooks devoted to teaching the various disciplines of Newtonian fluid dynamic ! This* is most *unfortunate, at least from my point of view, mainly because,* in the framework of a *university education, students obtain only a very "shallow" understanding of the real significance of the NSF system?* As a consequence, *for students* and also *certainly for many fluid dynamicists involved in research, fluid dynamics is a curious miscellany of different model problems which do not seem to bear any relation to each other, and particularly with the NSF system of equations, which is considered as a strange and even mystically unapproachable system.*

*The present view, developed in* Chaps. 6 and 8, *offers a way towards a full understanding of the origins of the more useful working models and gives an answer to the following question: do you know from whence it comes?* This *new view* also allows one to obtain *a classification of these working model problems emerging from various NSF initial-boundary value problems governing different fluid flows, and closely linked with the limiting values (zero or infinite) of reduced non-dimensional parameters in such problems.*

Another amazing *new perspective provided* by the RAM approach relates to the *reconsideration* of various *non-viscous* or *incompressible* reduced models via the corresponding NSF problems considered *as weakly viscous* or *slightly compressible* (see, for instance, Sect. 8.5, on the *classical Blasius BL problem).* Obviously, at the present time, this *desconstruction-RAM approach* and *reconsideration* is far from being an accomplished fact! But it is nevertheless clear that the origin of the equations for inviscid and incompressible fluid flows is the full system of unsteady NSF equations (3.2a–3.2d) given in Sect. 3.2, when Re $\rightarrow \infty$ and M $\rightarrow 0$!

But *unfortunately,* it should be noted that the limiting cases of steady and unsteady fluid flows from the steady or unsteady NSF system of equations leads to *very different classifications,* mainly because of the singularities in the space-time fluid flow domains during these two (simultaneous?) limiting processes. *In particular, near the initial time and in the vicinity of the wall of a body in motion, the above two limiting processes* $\mathrm{Re} \to \infty$ and/or $\mathrm{M} \to 0$ are very singular, because of the filtering of acoustic waves when $\mathrm{M} \to 0$ and modification of the mathematical nature of the equations (NSF to Euler) when $\mathrm{Re} \to \infty$.

Note that, in Fig. 5.1, some reduced approximate models (mentioned as *4-Useful limiting forms*) concern the RAM approach to the NSF problem and these models are also represented in Fig. 5.2, which gives a more detailed family of models resulting from the NSF equations for the cases when $\mathrm{Re} \gg 1$, $\mathrm{M} \ll 1$, and $\mathrm{Re} \ll 1$, the *isentropic NS viscous model (!)*—à la *Lions (1998)*—standing on its own, *isolated* on the right-hand side, aloof from the NSF model, assumed exact, without any descendant, and in fact wrong!

As a consequence of the singularities mentioned above, one must consider some *localized regions* in the space-time fluid flow domain, in particular *close to the initial time* $t=0$, for *atmospheric motions, where the initial conditions play a crucial role for the weather forecast, namely, prediction of the weather for the next few hours or longer*! The *main cause* of such a singularity is the *strong reduced form* of the *vertical equation of motion* in the *Prandtl unsteady boundary-layer system of model equations* for the case $\mathrm{Re} \gg 1$.

In Sect. 6.1, *we consider* an unsteady fluid flow at $\mathrm{Re} \gg 1$ using the RAM approach and the consequences of this reduction are considered. More precisely, for this unsteady fluid flow with vanishing viscosity, as a consequence of the constancy of the *pressure in the vertical* (relative to the z coordinate) in the boundary layer theory, the associated unsteady *Prandtl equations break down near the initial time*, where the initial data are given for the full unsteady atmospheric NSF equations.

*In an unsteady case, for high Reynolds numbers* $\mathit{Re} \gg 1$, *as was first shown in our short Note 1980* <1>, *we have in fact four regions:*

*The Rayleigh corner region (I)—close to* $t=0$ *and in the vicinity of the wall.*
*The acoustic non-viscous region (II)—close to* $t=0$, *but far from the wall.*
*The Prandtl BL region (IV)—far from the initial and wall regions.*
*The Euler non-viscous region (V)—far from the wall region.*

*But there is also an intermediate viscous matching region fifth (III) between the Prandtl BL and Rayleigh corner regions. This was also discovered in the Note (1980) mentioned above (see also* Fig. 6.1*).*

In particular, in the local Rayleigh-Howarth viscous and heat conducting region (the small corner region I), we recover the so-called "*compressible Rayleigh equations*" of *Howarth (1951)* as a significant local degeneracy of the full unsteady NSF equations near the initial time and close to the wall.

These regions I–V are *not mentioned* in Fig. 5.2, which shows only the classical *regular coupling* between the *Euler and Prandtl equations*, valid in fact only *in the steady case*, for fixed time t, in the above matching process (3.9a, b)–(3.9c). In Sect. 6.1, the reader can find a spectacular "*deconstruction scenario*" and also a sketch (Fig. 5.1) showing the structure of these four regions at $\mathrm{Re} \gg 1$, with an intermediate matching region III.

In the schematic diagram of Fig. 5.3, *the non-interacting coupling of* Euler *(inviscid)* and Prandtl *(BL) fluid flows for steady Re ≫ 1* is reproduced from p. 245 of our 2002 book, FMAI 64 [7], cited in "Introduction".

It is also interesting to observe that the local Rayleigh-Howarth viscous and heat conducting region (small corner region I) plays a crucial role in the unsteady case, when the function Θ in the thermal condition (2.10d) is assumed to be a function of time. Then, in dimensionless form

$$\mathrm{T} = 1 + \Lambda\{\Theta(\mathbf{P}) + \mathrm{Q}([\mathrm{t} - \mathrm{t}_{\mathrm{Ac}}]/(1/\mathrm{Re}))\} \text{ on the wall, with } \mathbf{P}(\mathrm{x},\ \mathrm{y}), \tag{5.1a}$$

where the *added function* Q is taken into account in the case of a thermal accident at the time $\mathrm{t} = \mathrm{t}_{\mathrm{Ac}}$, so that

$$[\mathrm{t} - \mathrm{t}_{\mathrm{Ac}}]/(1/\mathrm{Re}) = \theta. \tag{5.1b}$$

In the simulated problem, in the framework of the Rayleigh corner, the *time* is θ.

The unsteady singular character of the formulated typical NSF initial-boundary value problem then relates mainly to the sudden change in the wall temperature field at time $\mathrm{t} = \mathrm{t}_{\mathrm{Ac}}$, and then to the time, for some thermal evolution problem!

In Figs. 5.4 and 5.5 as a complement to Fig. 5.2, the reader can find various *steady working models* for Re ≫ 1, and also *unsteady working models* for Re ≪ 1, with some comments concerning these two limiting cases. Concerning the limiting case Re ≫ 1, the reader can find various asymptotic topics with Re ≫ 1 in our <FMIA 64>, published in 2002 [7], cited in "Introduction". For example, in pp. 215–292, deal with the theory of boundary layers, a concept which allows one to model fluid flows with vanishing viscosity near the wall of a solid body. Mathematically, the inviscid solution cannot be uniformly valid when the viscosity tends to zero or Re → ∞, because it fails to satisfy the no-slip boundary condition on the body. This non-uniformity *can only* be removed by introducing a *(thin)* boundary layer, according to *Prandtl (1904)*: near the body, viscosity matters. In fact, for $\nu^\circ = 0$, the Navier equation reduces to the Euler equation for an incompressible homogeneous and non-viscous fluid flow! This is also a *typical singular perturbation* problem!

Thanks to the Prandtl BL concept and the asymptotics (inner-matching-outer) theory, initiated by Kaplun, the vanishing viscosity limit is actually well understood by fluid dynamicists, but the mathematically rigorous theory (existence and uniqueness) is very difficult!

The reader can find a consistent derivation of "dominant equations" from the steady NSF equations, valid near the wall of a 3D solid body, which is the first step in the derivation of the 3D steady Prandtl BL equations. The process of matching with the Euler equations is discussed and the second order BL equations for a steady 2D flow are derived, and their influence on the inviscid outer fluid flow is considered.

But curiously, the BL equations are singular near the initial time t = 0, since a characteristic property of the boundary layer is the constancy of the pressure across the thickness of the BL. The initialization problem via the so-called Rayleigh-

Howarth equations is also discussed (see also Sect. 8.1). In BL theory the most familiar case is the basic *Blasius problem* for a steady (and *incompressible*) flow past a solid flat plate oriented in the direction of a uniform stream. This Blasius problem is also considered for a *slightly compressible* fluid flow in Sect. 8.5.

Finally, the Navier incompressible fluid flow with a variable viscosity is considered, and this leads to a three-layer asymptotic model. The modelling of a fluid flow within the Taylor shock layer is briefly investigated!

In his paper entitled *Singular Coupling and the Triple Deck Model*, Keith Stewartson attributed the discovery to Neiland (1969), Messiter (1970), and Stewartson and Williams (1969)! But, in reality, his own contributions (see Stuart (1986)) to the origins and development of their theory where truly remarkable. In Zeytounian, FMIA 64 [7] cited in "Introduction", Chap. 12 (2002), there are 54 pages (from 471 to 525) plus pp. 542–545 of references concerning various facets of the "triple-deck" theory!

Regarding the limiting case $\mathrm{Re} \ll 1$, in pp. 327–367 of <FMIA 64>, some aspects of the Stokes and Oseen models are considered, for steady and also unsteady incompressible Navier fluid flows, and also for steady compressible and heat conducting NSF fluid flows! Two limiting processes play a fundamental role, according to Lagerstrom (1964), in the asymptotic study of low-Reynolds number flow; namely the Stokes inner limit and the Oseen outer limit. But for a compressible flow when $Re \to 0$, one must specify the role of the *Mach number M*. In fact, one must pose the problem concerning the behaviour of the NSF equations when $\mathrm{Re} \to 0$ and $\mathrm{M} \to 0$ simultaneously.

Naturally, first of all, for the validity of the NSF system, it is obviously assumed that the limiting compressible flow, at low Reynold and low Mach numbers, remains a continuous medium, and this implies that the so-called dimensionless Knudsen number $\mathrm{Kn} = \mathrm{M/Re}$ is also a small parameter: $\mathrm{Kn} \ll 1$! As a consequence, the above double limiting process must be carried out with the following similarity relation:

$$\mathrm{M} = \mathrm{R}^* \, \mathrm{Re}^{1+a}, \text{ with } \mathrm{R}^* = \mathrm{O}(1) \text{ and } a > 0, \text{ when } \mathrm{Re} \to 0. \tag{5.2}$$

The case of the unsteady incompressible, low Reynolds number, Stokes problem in a bounded domain $\Omega(\mathrm{t})$ is also investigated. The starting problem is the Navier equations with the initial condition for the velocity $\mathbf{u}$ at $\mathrm{t}=0$, and a boundary condition on $\partial\Omega(\mathrm{t}) = \Sigma(\mathrm{t})$ under the Stokes limiting process:

$$\mathrm{Lim}^{\mathrm{Stokes}} = [\mathrm{Re} \to 0, \text{ with t and } \mathbf{x} \text{ fixed}]. \tag{5.3a}$$

Unfortunately, (5.3a) is singular (!) near $\mathrm{t}=0$. A process of adjustment to the initial condition at $\mathrm{t}=0$ is needed, with $\tau = \mathrm{t/Re}$, and the following limit:

$$\mathrm{Lim}^{\mathrm{Adj}} = [\mathrm{Re} \to 0, \text{ with } \tau \text{ and } \mathbf{x} \text{ fixed}]. \tag{5.3b}$$

The solution to this Stokes problem is derived in pp. 335–338 of FMIA 64.

Again, in the case of the exterior unsteady problem—flow past a sphere, for example—low Reynolds number flow is singular near the initial time and at infinity! Therefore, one must consider a three-region—initial, Stokes, and Oseen—unsteady asymptotic theory!

Finally, we consider a simple case of a horizontal layer of incompressible strongly viscous fluid, which moves between a fixed flat plane $z = 0$ and a free surface given by h(t, x, y), and via the "*long-wave approximation*", i.e., the horizontal wavelength $\lambda >> d$, where d is the thickness of the fluid layer when the fluid is at rest and the layer is bounded by an upper surface $z = d$.

The unsteady case of low-Reynolds number was considered by Bentwich and Miloh in various papers (1978, 1982, and 1984).

## 5.3 Some Remarks

The RAM approach was imagined mainly as an assistance to numericians confronted with various stiff computational problems with technological and geophysical fluid flows. What is aerodynamics, according to <NASA Knows! (Grades K-4)>: aerodynamics is the way air moves around things. The rules of aerodynamics explain how an airplane is able to fly. Anything that moves through air reacts to aerodynamics. A rocke*t blasting off the launch pad and a kite in the sky react to aerodynamics. Aerodynamics even acts on cars, since air flows around cars.*

On the other hand, it is well known that weather forecasting *is a crucial problem for many reasons in our daily activities, business, and pleasure,* with the burning question: *What will the weather be like tomorrow or for the next few days!*

For me it is obvious that our RAM approach is an indispensable tool, when linked with super-high-power computers and numerical analysis in the hands of numericians, for the simulation/computation of various problems! It is currently fashionable *to speak about the significance of numerical simulation (!),* thanks to the increased *capacity of supercomputers*, while no mention is made of the *role of fluid dynamics modelling for simulation with the help of a consistent rational working model*! Reference is rarely made *to the nature of the selected reduced computational model and its relation with the original mathematical formulation of the real simulated problem*!

The process of fluid dynamics modelling as an aid to numerical simulation and computation *does not* seem to *have any importance for many numericians!* In this situation, it is clearly necessary to ask an obvious question, one which requires serious thought, concerning these attractively presented numerical results: *How sure are you of your results? Indeed, all these numerical simulations/computations will cost money, and computerized numerical simulation, using a mathematical model created artificially in order to study what may exist in a real and very complicated fluid flow, is a very expensive activity, so confidence is necessary concerning the consistency of the working model and the abilities of this model as a valuable substitute for "reality"!*

Unfortunately, the models derived and used by numericians and applied mathematicians using ad hoc *procedures* will obviously not often *have* the *essential qualities* mentioned above! The consequences of this situation are the appearance of internal inconsistencies in most of the relevant intensive engineering computations based on these relatively ad hoc models! *Here, it seems to me once again very opportune to mention the extension of higher (at least second) order boundary layer theory to compressible and heat conducting flows, as carried out by Van Dyke (1962).* This consistent rational working model—in contrast to various ad hoc models proposed by consulting engineers (!)—was derived from the full NSF problem, *and constituted an important scientific contribution* to the NASA program for re-entry of the space shuttle from the space station,!

*In France, mathematics, and indeed mathematicians, holds a place of honour. Many of these mathematicians apply nonlinear abstract functional analysis in the hope of proving rigorous theorems about the existence and uniqueness of solutions for various fluid dynamics equations!* This is certainly an interesting and important field of research, but unfortunately, a careful inspection of various cases among their starting *fluid dynamics problems* shows that those problems are based on relatively ad hoc and highly simplified approximate reduced models, often completely lacking in physically realistic initial and boundary conditions and rife with internal inconsistencies.

This is particularly the case, as far as I know, in work relating to the *incompressible limit* of *compressible fluid flows* and in various supposedly rigorous papers devoted to *models for oceanic and atmospheric motions*! Again, the question is: *What is the scientific value of such rigorous results based on inconsistent* ad hoc *working models? In fact, it seems to me that the main objective of such investigations is mainly (and in many cases, only!) to demonstrate the abilities of their rigorous approach based on abstract nonlinear functional analysis—the considered model being merely a pretext!*

Perhaps the above verdict seems rather pessimistic, but I hope that brilliant young mathemathicians who feel somewhat attracted by fluid dynamics modelling using the RAM approach—rather than by abstract functional analysis—will be ready to work in this way. *For me, it is obvious that one should not underestimate the very substantial aid that the RAM approach may provide to the understanding of either the underlying physics or the mathematical structure of the working models under investigation. Indeed, the various applications of the RAM approach in* Chaps. 6 and 8 *provide clear confirmation of the above opinion!*

## Reference

<1> Zeytounian RKh (1980) Sur une forme limite des équations de Navier-Stokes, à grand nombre de Reynolds, au voisinage d'une paroi et de l'instant initial. CR Academic Science, Paris, 290, Série A, 567–569

# Chapter 6
# Key Applications of the RAM Deconstruction Approach

Here we present the deconstruction process via the RAM approach. Among the various examples, I have selected four of the most representative cases:

First, an unsteady aerodynamics problem with $Re \gg 1$ (see Fig. 6.1), with five regions already mentioned above.

Then, the unsteady fluid flow case with $M \ll 1$, and the filtering process of acoustic waves, which leads to approximate Navier-Fourier model equations (see Fig. 6.2).

In Figs. 6.1 and 6.2, the singularities are well illustrated by various localized fluid flow models, close to the initial time and near the wall, while Fig. 6.3 sketches some meteo-fluid dynamics models, illustrating the wide variety of working atmospheric models, as expounded in our 1990 monograph [22], cited in "Introduction".

Finally, Fig. 6.5 shows the three approximate models emerging from the classical Bénard problem via the RAM approach, when the liquid heated from below is weakly expansible, while the Fig. 6.4 shows the geometry of the full Bénard convection problem.

In Chaps. 2–12 of our monograph 2002, FMIA 64 [7], cited in "Introduction", the reader will find an application of the RAM approach to various fluid flow phenomena, arising from the full unsteady NSF initial-boundary value problem. These models are sketched in Fig. 5.1 of preceding Chap. 5.

Concerning the meteo-fluid dynamics models, the reader is invited turn to [21 and 22], cited in "Introduction", and the book published in 1991 <3>, cited in Chap. 2. But, in fact, my adventure with the modelling of atmospheric motions began in Moscow in the years 1957–1966, with I.A. Kibel, in the Department of Dynamic Meteorology of the Academy of Sciences, with my research on local winds over a thermally inhomogeneous mountain slope, and lee wave motion above and downstream of a mountain. The curious reader will find more information about my Moscow adventure with I.A. Kibel in my "Scientific Autobiography", published by Springer in 2014.

In France, on my return from Moscow (September 1966) and after the defense (in 1969) of my thesis for the Doctorat d'État ès Sciences Physiques, my first success, thanks to Paul Germain, who forced me (!) to *justify the Boussinesq*



R.Kh. Zeytounian, *Challenges in Fluid Dynamics*,
DOI 10.1007/978-3-319-31619-2_6

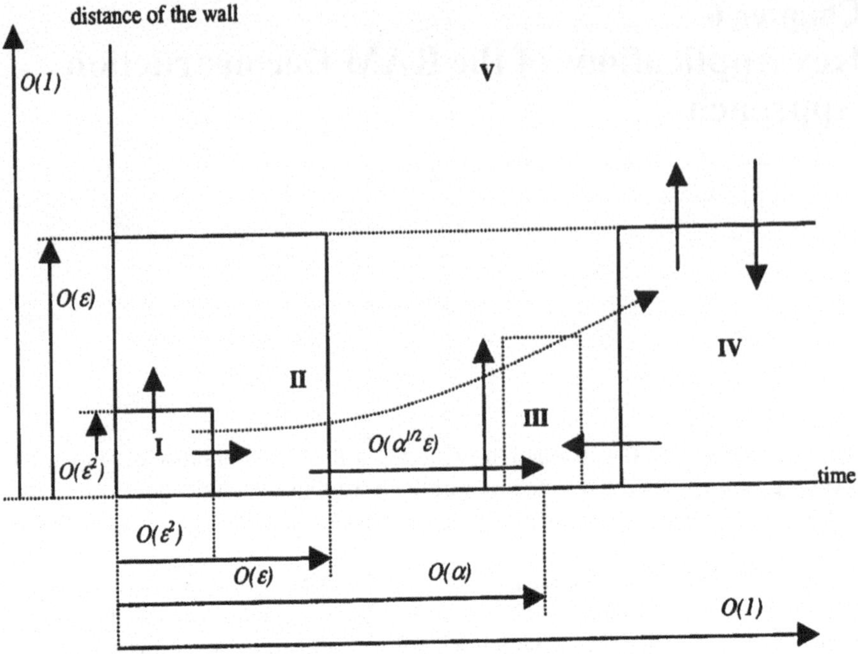


**Fig. 6.1** Four region structure with an intermediate matching region

*approximation* via *asymptotics*, was my 1974 paper <4>, cited in Chap. 3, submitted in Gdansk (Poland, 1973), in the XI Symposium on Advanced Problems in Fluid Dynamics. In the XII Biennial Fluid Dynamics Symposium in Bialowieza (Poland, 1975), I was invited to submit a general lecture entitled *La Météorologie du point de vue du mécanicien des fluides*, 1976 [21], cited in "Introduction". In English Summary, I wrote:

> The concept of model in dynamic meteorology is a fundamental one in view of the complexity of problems to be solved. However, in most cases, these models are built in a non-rational, intuitive manner, so it is difficult to insert the model within a system of rational approximations, which are derived from the general equations of fluid mechanics (written in a meteorological context!).

The reader can find the references to the papers mentioned above in our Scientific Autobiography of 2014.

We also observe once again that the application of the RAM approach to Bénard's famous problem, and the consideration of a weakly expansible liquid heated from below, was mainly inspired by reading the book *Physical Hydrodynamics* by Guyon, Hulin, Petit, and Mitescu (Oxford University Press, Oxford, 2001) <6>, cited in Chap. 3. In particular, the unexpected aspect of a weakly expansible liquid was surprisingly the appearance of an expansion parameter α. The key was to rewrite the Grashof number Gr as a ratio of two parameters (see Chap. 1 of [25], cited in "Introduction"), i.e., $Gr = \alpha/Fr^2$, in (6.17c) with (6.15b,c) (see Sect. 6.4).

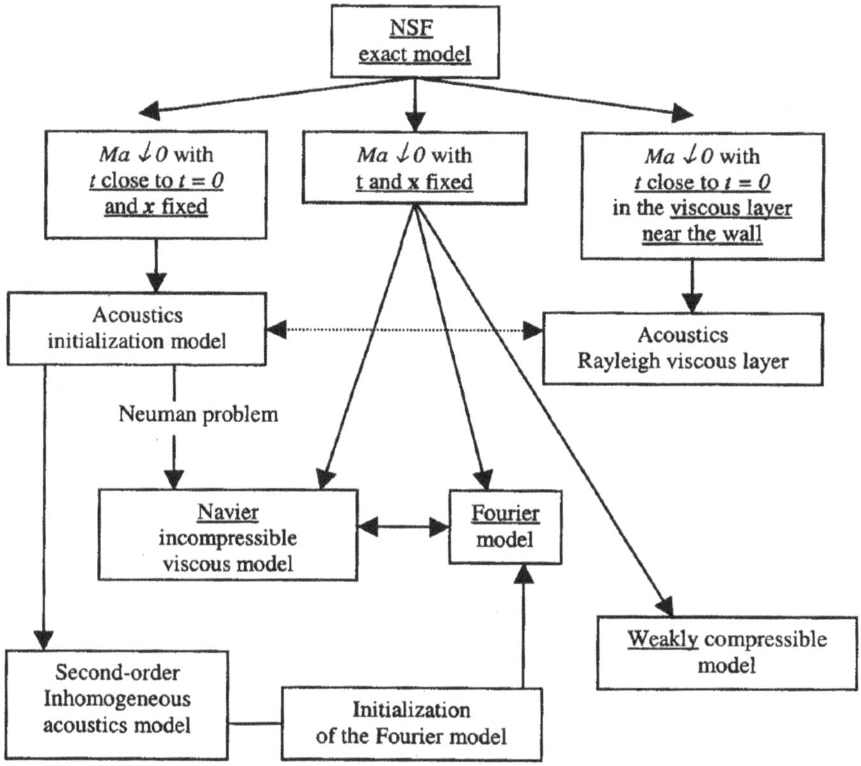


**Fig. 6.2** Navier-Fourier viscous incompressible model

## 6.1 The Unsteady High Reynolds Number Case (Re ≫ 1), and Fig. 6.1

It is well known that, when $\mathrm{Re}\uparrow\infty$ (or $\varepsilon = 1/\sqrt{\mathrm{Re}} \downarrow 0$), with t, x, y, and z fixed, as a consequence of the limiting process $\mathrm{Lim}^{\mathrm{E}}$ according to (3.9a), which leads from the NSF equations (3.3a–3.3d) to the degenerate non-viscous adiabatic Euler equations, one must consider, again from the NSF equations (3.3a–3.3d), the companion limiting process $\mathrm{Lim}^{\mathrm{Pr}}$: $\varepsilon\downarrow 0$ with t, x, y, $\zeta = z/\varepsilon$ fixed, according to (3.9b).

In MAE, for $\varepsilon \ll 1$, in the case of regular coupling (only actually valid in the steady case), we consider two asymptotic expansions:

$$(\mathbf{v},\ w,\ p, \rho, T) = (\mathbf{v}_E,\ w_E,\ p_E,\ \rho_E,\ T_E) + \varepsilon\left(\mathbf{v}^1{}_E,\ w^1{}_E,\ p^1{}_E, \rho^1{}_E,\ T^1{}_E\right) + \ldots, \tag{6.1a}$$

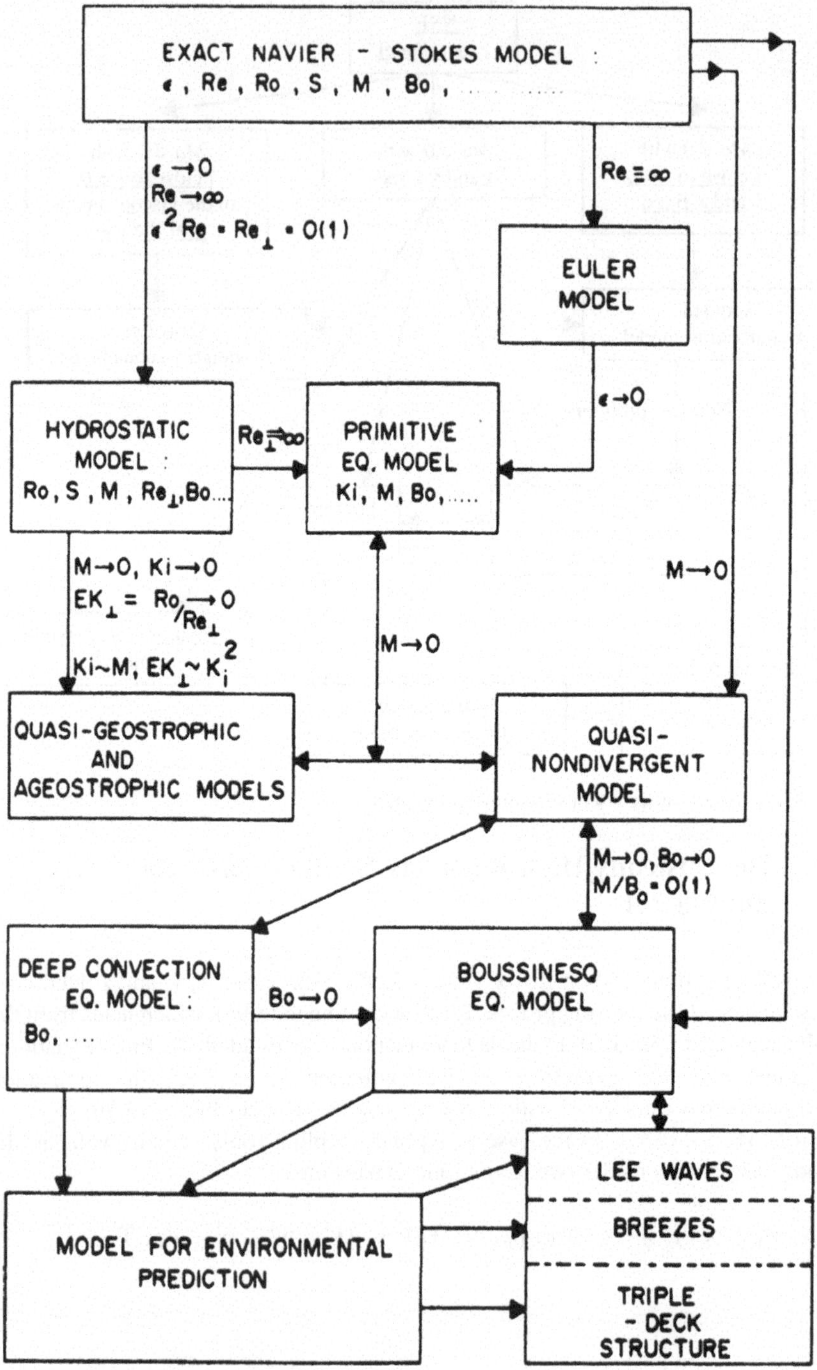


**Fig. 6.3** Meteo-fluid dynamics models

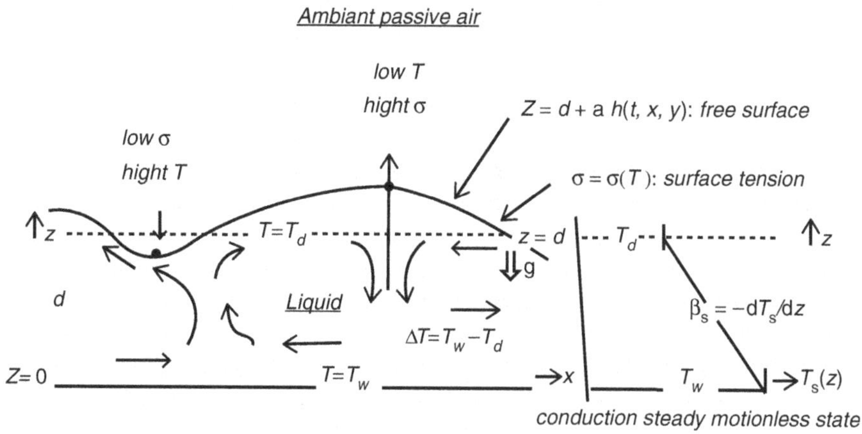


**Fig. 6.4** Geometry of the full Bénard convection problem

$$(\mathbf{v},\ w,\ p,\ \rho,\ T) = (\mathbf{v}_{Pr},\ 0,\ p_{Pr},\ \rho_{Pr},\ T_{Pr}) + \varepsilon\left(\mathbf{v}^1{}_{Pr}, w^1{}_{Pr},\ p^1{}_{Pr},\ \rho^1{}_{Pr},\ T^1{}_{Pr}\right) + \ldots. \tag{6.1b}$$

But the resulting unsteady Prandtl boundary equations are highly degenerate, since in place of a full unsteady equation for the vertical BL velocity component ($w^1{}_{Pr}$), we obtain, to leading-order, only the truncated equation

$$\partial p_{Pr}/\partial\zeta = 0 \Rightarrow p_{Pr} = p_{E|z=0} \text{ and } w_{E|z=0} = 0, \tag{6.2a, b, c}$$

where the term St $Dw^1{}_{Pr}/Dt$ is absent!

*As a consequence of* (6.2a), *the Prandtl system of boundary layer equations is, in fact, a consistent system of just two equations for the horizontal BL velocity vector* $\mathbf{v}_{Pr} = (u_{Pr}, v_{Pr})$ *and temperature* $T_{pr}$, *when we take into account the BL continuity equation*

$$\partial\rho_{Pr}/\partial t + \mathbf{D}.(\rho_{Pr}\mathbf{v}_{Pr}) + \partial\left(\rho_{Pr}w^1{}_{Pr}\right)/\partial\zeta = 0, \tag{6.3a}$$

in the Prandtl system of equations, with the equation of state

$$p_{E|z=0} = \rho_{Pr}T_{Pr}, \tag{6.3b}$$

*since* $w^1{}_{Pr}$ *must be computed using the following equation (assuming* $St \equiv 1$*):*

$$w^1{}_{Pr} = -(1/\rho_{Pr})\int^{\zeta}\{\partial\rho_{Pr}/\partial t + \mathbf{D}.(\rho_{Pr}\mathbf{v}_{Pr})\}d\zeta, \tag{6.3c}$$

with $\rho_{Pr} = p_E|_{z=0}/T_{pr}$, where $w^1{}_{Pr} = 0$ on $\zeta = 0$ and $p_E|_{z=0}$ is the Eulerian pressure at $z = 0$, a function only of the time t and x and y, which is related via the matching relation (3.9c) to $p_{Pr}$, and $\mathbf{D} = (\partial/\partial x, \partial/\partial y)$ is the horizontal gradient.

*However, curiously, the problem of two initial conditions for the two unsteady Prandtl boundary equations for* $\mathbf{v}_{Pr}$ *and* $T_{pr}$ *remains an open question even today, and it seems to me that a large majority of fluid dynamicists are not aware of this fact!* The strong singularity of the BL Prandtl equations near the initial time, where the initial data for the full unsteady NSF equations are given [see, for instance (2. 10a)], is just a direct consequence *of the change in the nature of the Prandtl equations* relative to the *incomplete parabolic character of the full unsteady NSF equations* (3.3a–3.3c).

In fact, for $\mathbf{v}_{Pr}$ and $T_{Pr}$, we have a system of two unsteady hyperbolic-parabolic equations instead of a *fourth order in time* system of unsteady NSF PDEs! This drastic change in the mathematical character of the full unsteady NSF system (3.2a–3.2c), which leads only to an unsteady Prandtl boundary layer, and a reduced system of two equations for $\mathbf{v}_{Pr}$ and $T_{Pr}$, brings about a significant modification which gives the Prandtl BL a singular nature near the initial time $t = 0$! It seems astonishing that such a singularity was ignored up to 1980; see, for instance, my short Note, 1980 <1>, cited in Chap. 5, where I wrote:

> In the present Note we show that the limiting form of the unsteady NSF equations for the compressible, slightly viscous and heat conducting fluid, at high Reynolds number, near the initial moment and in the vicinity of the wall of a body in motion, is identified with the classical Rayleigh equations (in fact, a significant degeneracy of the NSF equations when $Re\uparrow\infty$, but with $\theta = t/\varepsilon^2$ and $\eta = z/\varepsilon^2$ fixed, where $\varepsilon^2 = 1/Re!$). The process of matching between the Rayleigh equations and the Prandtl unsteady boundary layer equations is also described (see Fig. 6.1), and this permits us to elucidate the conditions under which the process of initialization of this Prandtl unsteady boundary layer must be made.

In particular, in Van Dyke's discussion of the nineteenth-century roots of the boundary layer idea (1994), this singular behaviour of the BL near $t = 0$ is completely overlooked!

> This new complicated structure, for unsteady viscous compressible heat conducting fluid flow equations at high Reynolds numbers seems in particular, very important for various applications linked to heat emergency-accident situations (the sudden rise of a thermal source, explosions, fires, failures at oil and gas pipelines, forest fires, etc.) in a local region on the wall in contact with the atmospheric air, at the fixed accident time $t_A$!

*Now, concerning the problem of prescribing two initial conditions for* $\mathbf{v}_{Pr}$ and $T_{pr}$, *for the two unsteady Prandtl BL equations, at* $t = 0$, *these will certainly differ (see, for instance, the unsteady adjustment to hydrostatic balance in* Sect. 3.4*) from the given initial data for the full unsteady starting NSF equations, maybe because the fluid flow situation is very different!* It should also be noted that, in 1912, Hilbert also encountered this singularity problem; see Chap. 1 in my book, 2002 [13], cited in "Introduction", and also the famous paper by *Grad (1969)* entitled *Singular and nonuniform limits of solutions of the Boltzmann equation.*

The *main question here is this*: what *are the two initial unknown conditions for the two unsteady Prandtl BL equations in region IV, and how are these initial conditions linked with the prescribed data at* $t = 0$, *for the full unsteady NSF system?* The RAM approach, via an adjustment problem, shows that the answer is closely related to the determination, from the full unsteady NSF equations (3.23–3.23d), of two particular equations valid near the initial time, but written with two different short times, namely, $\tau = t/\varepsilon$ and $\theta = t/\varepsilon^2 \equiv \tau/\varepsilon$.

In fact, we also have an *intermediate time*, for the intermediate matching region III, between the regions I (Rayleigh) and IV (Prandtl BL) (see Fig. 6.1). The *acoustic region II* is characterized by the following limiting process, when $\varepsilon \downarrow 0$:

$$\mathrm{Lim}^{\mathrm{Ac}}(\tau = t/\varepsilon, \text{ and } x,\ y, \zeta = z/\varepsilon \text{ fixed}). \tag{6.4a}$$

To leading order, we then derive for $w_{Ac}$, $p_{Ac}$, $\rho_{Ac}$, and $T_{ac}$ the usual classical non-viscous acoustic equations for an unsteady one-dimensional vertical $(\tau, \zeta)$ motion of the non-viscous adiabatic gas dynamics, and a 'transport' equation for the horizontal component $\mathbf{v}_{Ac}$ of the velocity vector.

For these gas dynamics equations, we write only a slip boundary condition

$$w_{Ac} = 0 \text{ on } \zeta = 0, \text{ when } \tau > 0, \tag{6.4b}$$

and also the above matching (in time) condition:

$$\lim_{\tau \uparrow \infty}\{\mathrm{Lim}^{\mathrm{Ac}}(\tau, \text{ and } x,\ y, \zeta \text{ fixed}) = \lim_{t=0}\{\mathrm{Lim}^{\mathrm{Pr}}(t, \text{ and } x,\ y, \zeta \text{ fixed})\}. \tag{6.4c}$$

It should be stressed that the aim of the unsteady adjustment problem is to clarify *just how a set of initial data associated with an exact starting system of equations (the unsteady NSF system) can be related to another set of initial data associated with an approximate reduced model system (Prandtl BL equations), which is a significant degeneracy of the original NSF starting system of equations.*

*When* $\tau \to \infty$, *the unsteady adjustment problem makes it possible via matching (if we do actually have a tendency towards a limiting state) to derive suitable initial conditions (at* $t = 0$*) for the approximate, unsteady time evolution, BL equations derived from the NSF system for fixed t!*

*But with the second short time*

$$\theta \equiv \zeta/\varepsilon = z/(1/Re), 1/Re = \varepsilon^2, \tag{6.4d}$$

*near the initial time, we also derive, from the full unsteady NSF equations* (3.2a–3.2c), *a new approximate reduced system of equations that is significant near the initial time and close to the wall, in a corner Rayleigh region I. In this case, with a Rayleigh limiting process*, where $\eta = z/\varepsilon^2$,

$$\mathrm{Lim}^{\mathrm{Ra}}(\theta, \text{ and } x,\ y, \eta \text{ fixed}), \tag{6.5a}$$

such that

$$\lim_{\theta\to\infty}\left[\mathrm{Lim}^{\mathrm{Ra}}\right] = \left[\mathrm{Lim}^{\mathrm{Pr}}\right]_{t=0}, \tag{6.5b}$$

*a matching condition between the BL equations with t and ζ fixed, with the local in time Rayleigh equations near θ = 0, and in the vicinity of the wall.* The model equations for the corner region I are those used in the Rayleigh compressible problem by Howarth (1951), but in the case of an infinite flat horizontal plate (in a viscous and heat conducting, originally quiescent fluid), which is impulsively made to start moving in its own plane with a constant velocity.

Figure 6.1 represents the four region structure with an intermediate matching region III. This new matching region III appears when we investigate first the far behaviour of the Rayleigh equations, valid in region I, for large values of θ and η, via the intermediate variables t* and z*, and write

$$t^* = \kappa(\varepsilon)\theta \text{ and } z^* = [\kappa(\varepsilon)]^{1/2}\eta, \text{ with } w_{\mathrm{In}} = w_{\mathrm{Ra}}/\kappa(\varepsilon), \tag{6.6a, b}$$

*when* κ(ε) tends to zero, *for fixed* t* and z* *in region III, and variables θ and η from the Rayleigh region I tend to infinity. Then*

$$\lim_{z^*\uparrow\infty}\left\{\mathrm{Lim}^{\mathrm{In}}\right\} = \lim_{\eta\downarrow 0}\left\{\mathrm{Lim}^{\mathrm{Ra}}\right\}, \tag{6.6c}$$

where

$$\mathrm{Lim}^{\mathrm{In}} = \left[t^* = t/\alpha(\varepsilon) \text{ and } x,\ y,\ z^* = z/\varepsilon\,\alpha(\varepsilon)^{1/2} \text{fixed})\right], \tag{6.6d}$$

*the intermediate variables being directly related to the initial dimensionless variables t and z, for the NSF equations* (3.2a–3.2c).

*As a general rule, the intermediate gauge* α(ε) *in* $\mathit{Lim}^{In}$ (6.6d) *differs from* κ(ε). *These two gauges both* ↓0, *when* ε↓0, *and a simple investigation shows that*

$$\alpha(\varepsilon) = \varepsilon^2/\kappa(\varepsilon). \tag{6.7a}$$

*This is a compatibility relation, and a direct consequence of the behaviour of the Prandtl BL equations, valid in region IV, when t and ζ both tend to zero, towards the intermediate region III.* Indeed, with t* and z* fixed, if

$$t^* = t/\alpha(\varepsilon) \text{ and } z^* = \zeta/\alpha(\varepsilon)^{1/2}, \text{ with } w_{\mathrm{In}} = \alpha(\varepsilon)w_{\mathrm{Pr}}, \tag{6.7b, c}$$

when α(ε)↓0, *then we once again derive the same, newly discovered, intermediate-matching model equations via the Prandtl BL equations. In particular, the intermediate matching model equations are derived in the following form:*

$$\partial\rho_{\mathrm{In}}/\partial t^* + \partial(\rho_{\mathrm{In}}w_{In})/\partial z^* = 0, \tag{6.8a}$$

$$\partial p_{In}/\partial z^* = 0, \quad p_{In} = \rho_{In} T_{In}, \tag{6.8b, c}$$

$$\rho_{In}[\partial T_{In}/\partial t^* + w_{In}\partial T_{In}/\partial z^*] + (\gamma - 1)p_{In}\partial w_{In}/\partial z^* = (\gamma/Pr)\partial^2 T_{In}/\partial z^{*2} + \gamma(\gamma - 1)M^2 \mid \partial \mathbf{v}_{In}/\partial z^*|^2, \tag{6.8d}$$

$$\rho_{In}[\partial \mathbf{v}_{In}/\partial t^* + w_{In}\partial \mathbf{v}_{In}/\partial z^*] = \partial^2 \mathbf{v}_{In}/\partial z^{*2}. \tag{6.8e}$$

Regions I–IV are sketched in Fig. 6.1. Unfortunately, a *precise localization* of the above intermediate matching equations (6.8a–6.8e), valid in region III, between the regions I and IV, does not seem possible at this stage of asymptotic analysis! More careful (second order) investigations are obviously necessary.

*Region I is the Rayleigh viscous corner region (close to time = 0 and close to the wall), where the equations for the unknown functions* $\mathbf{v}_{Ra}$, $w_{Ra}$, $p_{Ra}$, $\rho_{Ra}$, *and* $T_{ra}$, which depend on the time $\theta = t/\varepsilon^2 \equiv \tau/\varepsilon$ and the vertical coordinate $\eta = z/\varepsilon^2$, are:

$$\partial \rho_{Ra}/\partial \theta + \partial(\rho_{Ra} w_{Ra})/\partial \eta = 0, \tag{6.9a}$$

$$\rho_{Ra}[\partial w_{Ra}/\partial \theta + w_{Ra}\partial w_{Ra}/\partial \eta] + (1/M^2\gamma)\partial p_{Ra}/\partial \eta = (4/3)\partial^2 w_{Ra}/\partial \eta^2, \tag{6.9b}$$

$$\rho_{Ra}[\partial T_{Ra}/\partial \theta + w_{Ra}\partial T_{Ra}/\partial \eta] + (\gamma - 1)p_{Ra}\partial w_{Ra}/\partial \eta = (\gamma/Pr)\partial^2 T_{Ra}/\partial \eta^2 + \gamma(\gamma - 1)M^2\{\mid \partial \mathbf{v}_{Ra}/\partial \eta|^2 + (4/3) \mid \partial w_{Ra\ R}/\partial \eta|^2\}, \tag{6.9c}$$

$$\rho_{Ra}\ [\partial \mathbf{v}_{Ra}/\partial \theta + w_{Ra}\partial \mathbf{v}_{Ra}/\partial \eta] = \partial^2 \mathbf{v}_{Ra}/\partial \eta^2, \tag{6.9d}$$

with $p_{Ra} = \rho_{Ra} T_{Ra}$.

*Region II is the acoustic non-viscous region (close to time = 0, but far from the wall), where the equations for the unknown functions* $\mathbf{v}_{Ac}$, $w_{Ac}$, $p_{Ac}$, $\rho_{Ac}$, *and* $T_{Ac}$, which depend on the time $\tau = t/\varepsilon$ and the vertical coordinate $\zeta = = z/\varepsilon$, are:

$$\partial \rho_{Ac}/\partial \tau + \partial(\rho_{Ac} w_{Ac})/\partial \zeta = 0, \tag{6.10a}$$

$$\rho_{Ac}[\partial w_{Ac}/\partial \tau + w_{Ac}\partial w_{Ac}/\partial \zeta] + (1/M^2\gamma)\partial p_{Ac}/\partial \zeta = 0, \tag{6.10b}$$

$$\rho_{Ac}[\partial T_{Ac}/\partial \tau + w_{Ac}\partial T_{Ac}/\partial \zeta] + (\gamma - 1)p_{Ac}\partial w_{Ac}/\partial \zeta, \tag{6.10c}$$

$$\rho_{Ac}[\partial v_{Ac}/\partial \tau + w_{Ac}\partial v_{Ac}/\partial \zeta] = 0. \tag{6.10d}$$

*Region III is the intermediate viscous matching region (between the Prandtl BL and the Rayleigh corner regions), where we have the above system of equations* (6.8a–6.8e) for the *unknown functions* $\mathbf{v}_{In}$, $w_{In}$, $p_{In}$, $\rho_{In,}$ *and* $T_{In}$, which depend on the time t* and the vertical coordinate z*, given by (6.7b,c) with (6.7a).

*Region IV is the Prandtl BL region (far from the initial and wall regions), where the equations for the unknown functions* $\mathbf{v}_{Pr}$, $w_{Pr}$, $p_{Pr}$, $\rho_{Pr}$, *and* $T_{Pr}$, which depend on the time t and the vertical coordinate $\zeta = = z/\varepsilon$, and also on x and y, with $\mathbf{D} = (\partial/\partial x, \partial/\partial y)$, are the full system of BL equations.

*Region V is the Euler non-viscous region (far from the wall region), where we have the classical non-viscous equations.*

We also observe that, when $\alpha(\varepsilon) = \varepsilon^2$, we recover the Rayleigh region I, and when $\alpha(\varepsilon) = 1$, we recover the Prandtl region IV. The existence of such an intermediate region is a striking indication that it seems possible (as a conjecture?) to *directly match the Rayleigh and Prandtl equations by an adjustment problem*, via the intermediate equations, when $t^*\uparrow\infty$, using the matching conditions (in time):

$$\lim_{t^*\uparrow\infty}[v_{In}, T_{in}] = [v_{Pr}, T_{pr}]_{t=0}. \tag{6.9}$$

Finally, we observe that it is indeed necessary *to complete* our above *investigations* with some detailed applications of *asymptotic expansions, whence we may construct a hierarchy with the model equations discovered above!*

## 6.2 Unsteady Low Mach Number Case (M ≪ 1), and Fig. 6.2

The case $M \ll 1$, which leads to the so-called Navier-Fourier viscous incompressible and heat conducting model—a less complicated system of equations than the full unsteady NSF system of equations—seems to me better suited to a rigorous investigation by "*pure*" *mathematicians.*

First, we consider the following outer (t and **x** fixed) Navier incompressible limit:

$$\mathrm{Lim}^N = [M \to 0 \text{ with t and } \mathbf{x} \text{ fixed}], \tag{6.10a}$$

of the full unsteady NSF equations (3.3a–3.3d), written in dimensionless form [using, for simplicity, the same notations for the dimensionless quantities and neglecting the gravity term proportional to Bo in (3.3a)]. We can assume, for example, that the dimensionless pressure p is given by the asymptotic expansion

$$p = p_o(t) + M^2 p_N(\mathbf{x}, t) + o(M^2). \tag{6.10b}$$

Scrutinizing the equations and trying more sophisticated expansion processes proves that this expansion (6.10b) is the most general asymptotic expansion for the pressure in the unsteady case, *possibly with a term* proportional to $M^3$. First, we ask how we can get information about the unknown function $p_o(t)$? When (6.10b) holds in a neighbourhood of infinity, in the flow domain, where the pressure is usually constant, we may set $p_o(t) = 1$. As a matter of fact, we can put any constant in place of unity, but this is only a matter of a convenient choice for the reference pressure $p_c$ introduced in Sect. 3.2.

This occurs in what is called "*external aerodynamics*", a field of intensive study. On other hand, in "*internal unsteady aerodynamics*", let us assume that a gas is contained in a bounded region $\Omega$ (t) with an impermeable but possibly deformable (in time t) wall, so that the gas in $\Omega$ (t) is a given function of time. *An obvious way to proceed now is to assume that the associated density and temperature approach definite limits $\rho_o(t)$, $T_o(t)$, such that, $p_o(t)=\rho_o(t)T_o(t)$, in the (outer) low Mach number limit process* (6.10a) *with* (6.10b).

*It is a very easy matter to obtain, from the dimensionless energy equation (written for a perfect gas, $p=\rho\ T$), an equation satisfied by $T_o=T_o\ (t)$, which holds provided that*

$$T_o/(\rho_o)^{\gamma-1} \text{is independent of time t,} \tag{6.10c}$$

and this is indeed the case, at least far from the boundary. By simply choosing $p_c$ and $T_c$, we may assume that this constant is equal to unity, and as a consequence of the conservation of global mass $m^\circ$ for the whole of the gas contained within the container, the relations (6.10c) and (6.10d) are realized:

$$p_o(t) = 1/V(t)^\gamma, \text{ with } V(t) = (\rho_o/m^\circ)|\Omega(t)|, \tag{6.10d}$$

with $|\Omega\ (t)|$ the volume of the bounded container.

So we have found our way out of the indeterminacy concerning the leading term in (6.10b). Of course, our argument relies on $T_o(t)$ being independent of space, and we have to address the adequacy of that.

### *6.2.1 Navier-Fourier Limit Equations*

Let us return to the dimensionless continuity and Navier-Stokes equations, (3.3b) and (3.3a), respectively, and assume that **u** goes to $\mathbf{u}_N$ (t, x), according to the outer Navier limit process (6.10a). We set

$$p_N/\gamma p_o(t) = \text{Lim}^N\{[p - p_o(t)]/\gamma M^2 p_o(t)\} = \pi_N, \tag{6.11a}$$

a fictitious pressure, and obtain the following modified Navier equations for the velocity vector $\mathbf{u}_N$ and the pressure perturbation $\pi_N$:

$$\nabla.\mathbf{u}_N = -(1/\rho_o(t))d\rho_o/dt, \tag{6.11b}$$

$$[\partial/\partial t + \mathbf{u}_N.\ \nabla]\mathbf{u}_N + \nabla\pi_N = (1/\text{Re})\ (\mu(T_o)/\rho_o)\Delta\mathbf{u}_N, \tag{6.11c}$$

with

$$\rho_o(t) = m^\circ\, V(t)/|\Omega(t)|;\; p_o(t) = [1/V(t)]^\gamma;\; T_o(t) = [V(t)]^{1-\gamma}. \tag{6.11d}$$

In the above system of two limit equations (6.11b and 6.11c), with (6.11d), the Strouhal number St, which characterizes the unsteady effect, is assumed to be unity for simplicity. However, in some cases, it is useful to consider the large Strouhal number case, and as an example, this arises in the derivation of the equations of acoustics [see the system of equations (6.14a) and also Fig. 6.2].

This is also the case for high Strouhal number $St \gg 1$, with $M \ll 1$, such that the following similarity relation is satisfied:

$$M\,St \equiv 1,\ \text{when}\ (1/St) \downarrow 0\ \text{and}\ M \downarrow 0, \tag{6.11e}$$

and in such a case it is not necessary to introduce $\tau = t/M$, the short adjustment time, as in the case considered in Sect. 6.2.2.

Equations (6.11b and 6.11c), with (6.11d), are a slight variant of the classical (dimensionless) Navier equations for a fluid with time dependent viscosity, in a motion that, rather than being divergenceless, is one with a divergence that is constant in space and variable in time. The usual set of Navier incompressible equations is obtained for a constant volume container (or for external aerodynamics, when the basic thermodynamic state at infinity is constant). In this case we can write, in place of (6.11b and 6.11c), the usual dimensionless Navier equation:

$$[\partial/\partial t + \mathbf{u}_N.\,\nabla]\mathbf{u}_N + \nabla\pi_N = (1/Re)\Delta\mathbf{u}_N, \tag{6.12a}$$

with

$$\nabla.\mathbf{u}_N = 0, \tag{6.12b}$$

where

$$\pi_N = [p - 1]/\gamma M^2. \tag{6.12c}$$

For a constant volume container or for external aerodynamics, from the dimensionless temperature equation (3.3d) written for a perfect gas, we associate with (6.12a and 6.12b) and (6.12c) the following linear non-homogeneous Fourier-type equation, where the perturbation of the pressure $\pi_N$ consistently plays a role in the leading-order equation for $T_N$:

$$\begin{aligned}&[\partial/\partial t + \mathbf{u}_N.\,\nabla]T_N - (1/Pr\,Re)\Delta T_N\\&= [(\gamma - 1)/\gamma][\partial/\partial t + \mathbf{u}_N.\,\nabla\pi_N] + (2/Re)\,Tr[\underline{\mathbf{D}}(\mathbf{u}_N)]^2,\end{aligned} \tag{6.12d}$$

with

$$T_N = [T - 1]/M^2. \tag{6.12e}$$

In equation (6.12d), which is a limiting form of the NSF-type equation (3.3c) for the temperature, $\underline{\mathbf{D}}(\mathbf{u}_N)$ is the second-order rate-of-deformation tensor for the velocity $\mathbf{u}_N$.

Equation (6.12d) for the perturbation of the temperature $T_N$ given by (6.12e) is the only consistent equation, associated with the Navier equation (6.12a) for ($\mathbf{u}_N$ and $\pi_N$), and incompressibility constraint (6.12b), as a consequence of the low Mach number outer limit (6.10a), when we assume, not only (6.12e), but also the following similarity condition, for the wall temperature parameter $\Lambda$ in condition (3.4b):

$$\Lambda = \Lambda^* M^2, \text{ with} \Lambda^* = O(1) \text{ fixed.} \tag{6.12f}$$

The system of two equations (6.12a and 6.12b), with (6.12c), and equation (6.12d), with (6.12e), and the constraint (6.12f), for the three functions $\mathbf{u}_N$, $\pi_N$, $T_N$, with the parameter $\Lambda$, is the so-called Navier-Fourier model system of equations, according to Zeytounian [24, Chap. 4], cited in "Introduction", for external aerodynamics. In fact, this Navier-Fourier model system is a well formulated and valuable approximate reduced model, and less stiff than the typical full unsteady I-BV NSF problem, (3.3a–3.3d) with (3.4a–3.4c), formulated in Chap. 3.

Unfortunately, because the unsteady fluid flow is considered with the incompressibility constraint (6.12b), this Navier-Fourier system of equations is singular near the initial time where the data are given, even as simply as in (3.4c)! Near $t=0$, instead of the Navier-Fourier model system, an acoustic model works, and matching is required between the N-F model and the acoustic model. But it seems that, for this N-F model problem, the boundary conditions (3.4a) and (3.4b) remain true, despite the fact that, in (3.4b), in some circumstances, it may be necessary to give a precise formulation for the time dependence in the wall temperature field $\Theta$. This happens when dealing with a technological or thermal accident, owing to the sudden ignition on the wall, e.g., initiating a fire!

### *6.2.2 Equations of Acoustics*

When we consider the above unsteady Navier-Fourier (N-F) model system of equations, derived via the Navier outer limit (6.10a), with fixed time t, and incompressibility constraint $\nabla \cdot \mathbf{u}_N = 0$, we must give the initial value for $\mathbf{u}_N$ and $T_N$:

$$t = 0 : \mathbf{u}_N = \mathbf{u}_N{}^\circ(x),\ T_N = T_N{}^\circ(x). \tag{6.13a, b}$$

However, the initial conditions for the density $\rho$ have been lost during the outer Navier limiting process (6.10a), and the initial velocity field $\mathbf{u}_N{}^\circ(x)$ for the Navier equation (6.12a) must satisfy the divergenceless relation $\nabla. \mathbf{u}_N{}^\circ(x) = 0$, where $\mathbf{u}_N{}^\circ(x) = \mathbf{u}_N(0, \mathbf{x})$.

Obviously, the above Navier-Fourier limit equatio*ns* (6.12a) *and* (6.12d) *are not valid near* $t = 0$, *mainly because the acoustic waves are filtered in the above outer N-F equations, and as a consequence one must consider a short time that is significant in the region near time* $t = 0$:

$$\tau = t/(M)^{\alpha} = O(1),\ M \to 0, \tag{6.13c}$$

with the inner limit

$$\mathrm{Lim}^{Ac} = [M \to 0 \text{ with } \tau \text{ and } \mathbf{x} \text{ fixed}], \tag{6.13d}$$

instead of the outer limit (6.10a). The inner limit (6.13d), with short time $\tau$ given by (6.13c), is well-suited to studying this transient behavior to the N-F system of equations ($\alpha$ is a positive scalar). In fact, for the case of external aerodynamics, we can take as (inner, in time) solution the asymptotic expansions, associated with (6.13c) and (6.13d),

$$\mathbf{u} = \mathbf{u}_{Ac} + .., \tag{6.13e}$$

$$p = 1 + M^{a}\, p_{Ac} + .., \tag{6.13f}$$

$$\rho = 1 + M^{b}\, \rho_{Ac} + .., \tag{6.13g}$$

$$T = 1 + M^{c}\, T_{Ac} + .., \tag{6.13h}$$

for the NSF equations (3.3a–3.3d) with, for instance, the conditions (3.4a,b), where a, b, c, in (6.13f–6.13h) will be specified below.

*In fact, this transient behaviour which connects the acoustic region near the initial time with the N-F outer equations, valid far from* $t = 0$, *is essentially characterized by the effect of compressibility dominating viscosity! Considering the local (in time) inner acoustic limiting process* (6.13d), *from the continuity equation* (3.3b), *we first derive, with* (6.13c) *and asymptotic expansions* (6.13e–6.13h), *the least degenerate limit equation when* $\alpha = 1$ *and* $b = 1$. *It is then easy to convince oneself that the more consistent (dimensionless) limit system emerging from the full NSF equations* (3.3a–3.3d) *near the initial time is the following linear acoustics system of equations, when we take into account also* $c = 1$:

$$\begin{aligned} &\partial \rho_{Ac}/\partial \tau + \nabla .\, \mathbf{u}_{Ac} = 0,\\ &\partial \mathbf{u}_{Ac}/\partial \tau + (1/\gamma)\nabla p_{Ac} = 0,\\ &\partial T_{Aa}/\partial \tau + (1 - \gamma)\nabla .\, \mathbf{u}_{Ac} = 0, \end{aligned} \tag{6.14a}$$

with

$$p_{Ac} = \rho_{Ac} + T_{Ac}. \tag{6.14b}$$

To determine the two unknown initial conditions $\mathbf{u}_N{}^\circ(x)$ and $T_N = T_N{}^\circ(x)$ in (6.13a,b), for the Navier equation (6.12a) for $\mathbf{u}_N$ with (6.12b); and the Fourier equation (6.12d) for $T_N$, one must consider a matching condition:

$$(\mathbf{u}_{Ac}, T_{Ac})\Big|_{\text{when}\,\tau\to\infty} \textit{tends to}: \ (\mathbf{u}_N, T_N)\Big|_{\text{at } t=0}, \tag{6.14c}$$

and here we note only that, in *Wilcox (1975)*, the reader can find a scattering theory which provides a way to analyze the behaviour of the above acoustics equations (6.14a,b) when $\tau \to \infty$.

This matching was carried out in *Zeytounian (2000)*. In a numerical simulation, one must consider the acoustics system of equations (6.14a) and (6.14b), with *starting initial conditions (at $\tau = 0$) and boundary conditions,* subject to the full unsteady NSF problem, at the same time, with the Navier equation (6.12a) and Fourier equation (6.12d), and $\nabla . \mathbf{u}_N = 0$. It seems that these coupled problems and the associated conditions constitute an interesting challenge for numericians, but it is not an easy one, and it requires a lot of care and attention!

The results of this numerical simulation in the framework of the acoustics problem, when the computation with the short time $\tau$ is performed over a sufficiently long (but finite!) period, leads to the two unknown initial conditions for $\mathbf{u}_N$ and $T_N$ (6.13a,b), solutions of (6.12a) and (6.12d).

## 6.3 Meteo-fluid Dynamics Models, and Fig. 6.3

In Fig. 6.3, the various meteo-fluid dynamics models are sketched on the basis of the so-called tangent models resulting from the NSF atmospheric equations used for weather prediction (see Sect. 3.4). In the monograph [28], cited in "Introduction", the reader can find a *large family* (see, for instance, the above Fig. 6.3) of approximate model equations emerging from the NSF system of atmospheric equations. *This family emerges from the NSF system via the RAM approach, and corresponds to the reduced atmospheric models listed below:*

*Hydrostatic* model: when $\varepsilon$ *and (1/Re) both*$\downarrow 0$ such that $\varepsilon^2\text{Re} \equiv \text{Re}_\perp = O(1)$).
*Boussinesq* model: when *M and Bo both* $\downarrow$ *0* such that $\text{Bo}/M = B^* = O(1)$,

such a model being valid only in a thin layer of $\approx 1$ km.

*Isochoric* model: $\gamma\uparrow\infty$ *and* $M\downarrow 0$ such that $\gamma\, M^2 = M^* = O(1)$.
*Deep convection* model: an *outer model* relative to *Boussinesq, which is an inner model.*
*Quasi-geostrophic* and *ageostrophic* models: for *low Kibel number* flow.
*Low Mach* number flow: for lee waves *above* and *downstream* of a *mountain.*
*Quasi-nondivergent* model: *local* and *regional scale* atmospheric flows, *lee waves, breezes, environmental* prediction models and *triple-deck* structure

It is true that the modelling of atmospheric motions/flows, especially via well-balanced approximate and consistent models for global, short-range, and local weather forecasting, involve vast and complicated applied mathematics! The 24-year period of my scientific activity at the University of Lille was devoted mainly to the modelling of fluid flow phenomena, and in this scientific endeavour, I was strongly influenced by my own conception of meteorology. In fact, doubly so, as an: "unconscious inheritance" *from my scientific period in Moscow from 1957 to 1966, with Ylya Afanas'evich Kibel*, and as a fluid dynamics discipline which is a privileged area for application of the RAM approach!

It is obvious that the *basis for application of the RAM approach* to *atmospheric motions* mainly concerns *low Mach number flows*, and as a consequence, the Mach number M is the fundamental small parameter in *meteo—atmospheric fluid dynamics*. But in reality, some intriguing features are found with low Mach number atmospheric flows, due to the interactions of gravitational and Coriolis forces with pressure forces.

As a consequence one encounters a number of fascinating problems, some of which have been discussed in my 1990 book [22]. But one cannot avoid a feeling of frustation, the reason being that we do not encounter any model playing as central a role as the one played by incompressible fluid flows considered in Chap. V of our FMIA 64 [7], cited in "Introduction" (pp. 87–112 and 529–531). This *feeling of frustation* seems linked with blocking (in the horizontal planes). This is overwhelming, and induces a number of unpleasant features which may *lead the sceptical reader to believe more in numerical codes than asymptotic modelling*!

Despite this, I quote once again a remark *by Huijun Yang from the University of Chicago (see SIAM Review 33, Dec. 1991, pp. 672–673), concerning my 1990* book [28], cited in "Introduction":

> Throughout the book, the reader can feel the strong influence of Soviet works on the author. However, the author does have his own character. The issues raised in the book, such as the initialisation (initial layer), boundary layer treatment, and well-posedness and ill-posedness of the system, are very important problems facing researchers today in atmospheric sciences and other related sciences. The reader will find some valuable information on these issues... The mathematically consistent treatment of the subject does give this book a unique place on the shelves of libraries and in the offices of researchers.
>
> This book is very different from recent books on the market, i.e., Holton (1979), Gill (1982), Haliner and Williams (1980), Pedlosky (1987), and Yang (1990). I recommend that researchers in atmospheric dynamics and numerical weather prediction read this book to have an alternative view of the derivation of atmospheric flow models.
>
> Researchers in theoretical fluid mechanics might also be interested to see how singular perturbation methods can be used in atmospheric sciences. ... as the author says in his preface: 'I am well aware that this book is very personal, one might even say impassioned'.

At the present time, I think that *atmospheric motions,* viewed as *geophysical fluid dynamics problems*, pose for our environment numerous very interesting and specific challenges, and in Sects. 8.6 and 8.7 the reader will find two applications of the RAM approach to *terrestrial/environmental* problems which should be of particular interest!

## 6.4 The Three Significant Models of the Bénard Problem and Figs. 6.4 and 6.5

This last section of Chap. 6 is devoted to three significant reduced working convection models, emerging from the NSF system of equations, for weakly dilatable liquids: *Rayleigh-Bénard shallow thermal convection* via à la *Oberbeck and Boussinesq*; *deep thermal convection* (à la *Zeytounian*), when the viscous dissipation effect is taken into account; and *Bénard-Marangoni thermocapillary convection,* with a deformable free surface effect.

In fact, from the RAM approach, we observe that it is sufficient to assume the availability of the following very simple approximate reduced *specifying* equation of state—*à la Rayleigh*—between the density ρ and the temperature T, for a weakly dilatable liquid:

$$\rho \approx \rho(\mathrm{T}), \tag{6.15a}$$

Below, the reader can find logically consistent and convincing arguments which confirm (6.15a), at least for the derivation of the significant leading-order working models for the three above-mentioned convection problems concerning a *weakly dilatable liquid*!

The *weakness of the dilatation of the liquid is* characterized by the *small (expansion/dilatation) parameter*

$$\alpha = \ni\Delta\mathrm{T} \ << \ 1, \tag{6.15b}$$

with $\Delta\mathrm{T} = \mathrm{T}_1 - \mathrm{T}^\circ > 0$, where, as in the case of the classical Bénard (1900) thermal convection problem between two parallel plates, $\mathrm{T}_1$ is the *constant temperature at the rigid lower plate* ($x_3 = 0$) and $\mathrm{T}^\circ$ *the constant temperature of the rigid upper plate* ($x_3 = d$). The thermal dilatation-expansion coefficient of the liquid (at $\mathrm{T} = \mathrm{T}^\circ$, usually very small) is denoted by

$$\ni = \ - \ [1/\rho(\mathrm{T}^\circ)](d\rho/d\mathrm{T})_{\mathrm{T}\ =\mathrm{T}^\circ}. \tag{6.15c}$$

The liquid layer thickness is *d*, in the *non-deformable* free surface basic *conduction state,* and the main objective below is *to show that d is not arbitrary, but should satisfy some conditions (in fact restrictions!) for the validity of the asymptotic derivation of the so-called Oberbeck–Boussinesq (O-B) model equations from the full exact system of unsteady NSF equations* (see Sect. 2.2) *governing the evolution of the weakly expansible liquid.*

In Fig. 6.4, the reader will find the schematic *geometry of the full Bénard convection problem*, heated (and bounded) from below by a rigid horizontal flat surface (the *lower plate*, z = 0) and above by passive air. The rigid horizontal flat surface z = d is the *upper plate* of the weakly dilatable/expansible l*iquid layer* in the steady motionless conduction state, as represented on the right-hand side of Fig. 6.4 *(which has been taken from p. 93 of* [25], cited in "Introduction").

For such a *weakly dilatable liquid*, two non-dimensional parameters play an important role in the RAM approach, viz., the *Froude* and *Boussinesq* reduced parameters:

$$\mathrm{Fr}_d = (\nu^\circ/\mathrm{d})/\sqrt{\mathrm{gd}}, \tag{6.16a}$$

$$\mathrm{Bo} = \mathrm{gd}/\mathrm{C}^\circ \Delta \mathrm{T}, \tag{6.16b}$$

where it is assumed for simplicity that, at $T = T^\circ$, the *kinematic viscosity* $\nu^\circ$ and the *specific heat* $C^\circ$ at constant volume are both constant. In such a case, for *dimensionless velocity components* $v_i$, we write

$$\mathrm{v_i} = \mathrm{u_i}/(\nu^\circ/\mathrm{d}),\ \mathrm{i} = 1,\ 2,\ 3, \tag{6.16c}$$

and introduce the dimensionless perturbation of the pressure $\pi$ and reduced temperature $\theta$ by

$$\pi = [\mathrm{p}/(\rho^\circ \mathrm{gd})] - (1 - \mathrm{x_3}/\mathrm{d}). \tag{6.16d}$$

and

$$\theta = (\mathrm{T} - \mathrm{T}^\circ)/\Delta \mathrm{T}, \tag{6.16e}$$

with $\rho^\circ = \rho(T^\circ)$.

The introduction of $\pi$ and $\theta$ by (6.16d and 6.16e) justifies the *approximate relation* (6.15a), viz., $\rho \approx \rho(T)$ *of Rayleigh (1916)*, as a reduced equation of state for a weakly expansible liquid in leading order. The dimensionless time-space variables are

$$\tau = \mathrm{t}/\left(\mathrm{d}^2/\nu^\circ\right) \tag{6.16f}$$

and

$$\mathrm{X_i} = \mathrm{x_i}/\mathrm{d}, \tag{6.16g}$$

and this gives the possibility of deriving a new system of dominant dimensionless equations that is valid with an error of the order $O(\alpha^2)$, as shown in our 2009 book [3, Chap. 2], cited in Chap. 1. This new system of dominant dimensionless equations for weakly expansible liquids is well suited to application of the RAM approach.

Now, from these dominant leading-order equations, when $\alpha \to 0$, it is obvious (after a short of reflexion) that, if we want to take into account the *buoyancy term* $(\alpha/(\mathrm{Fr}_d^2)\ \theta\ \delta_{i3}$ in the leading-order equation for $v_{i0}$, the first term of $v_i$, we should make the following asymptotic expansions:

$$\begin{aligned} v_i &= v_{i0} + Fr_d^2 v_{i1} + \ldots, \\ \pi &= Fr_d^2 \pi_1 + Fr_d^4 \pi_2 + \ldots, \\ \theta &= \theta_0 + Fr_d^2 \theta_1 + \ldots., \end{aligned} \tag{6.17a}$$

which are associated with the following O-B limiting process:

$$\alpha \to 0 \text{ and } Fr_d \to 0, \text{ with } \tau \text{ and } X_i \text{ fixed, such that} \tag{6.17b}$$

$$\alpha / Fr_d^2 = Gr = O(1), \tag{6.17c}$$

where the *Grashof number* Gr is a *similarity parameter* for the above O-B limiting model equations (6.18a), written below for $v_{i0}$ ($i = 1, 2, 3$), $\pi_1$ and $\theta_0$.

Equations (6.18a) are derived when we assume that $\tau$, $X_i$, Pr, Bo, $\nu^\circ$, $k^\circ$, and $C^\circ$ are fixed (of the order O(1)), when $Fr_d \to 0$ with (6.17a–6.17c) and

$$\begin{aligned} &\partial v_{i0}/\partial X_i = 0, \\ &D^\circ v_{i0}/D\tau + \partial \pi_1/\partial X_i - Gr\,\theta_0 \delta_{i3} = \Delta_3 v_{i0}, \\ &D^\circ \theta_0/D\tau = (1/Pr)\Delta_3 \theta_0, \end{aligned} \tag{6.18a}$$

where $Pr = C^\circ \nu^\circ \rho^\circ / k^\circ$ is the Prandtl number.

In the framework of the classical Bénard problem heated from below, the boundary conditions relative to $X_3$ for the O-B equations (6.18a) are

$$v_{i0} = 0 \text{ and } \theta_0 = 1 \text{ at } X_3 = 0, \; v_{i0} = \theta_0 = 0 \text{ at } X_3 = 1, \tag{6.18b}$$

and in (6.18a),

$$D^\circ/D\tau = \partial/\partial\tau + v_{i0}\partial/\partial X_i \text{ and } \delta_{33} = 1, \delta_{13} = \delta_{23} = 0. \tag{6.18c}$$

Concerning the *validity of the O-B equations* (6.18a), we note first that the O-B model equations are valid only for a weakly expansible liquid when the parameter $\alpha$ given by (6.15b) is *a small parameter*. For the usual liquid this is very well satisfied if the temperature difference $\Delta T$ is not too large.

A second condition is related to the assumption of a *low Froude number*, such that the *similarity relation* (6.17c), which determines $Gr = O(1)$, *is well satisfied*. In particular, *from the similarity relation* $\alpha/Fr_d^2 = 1$, we obtain the following strong condition on the thickness d of the fluid layer:

$$d >> \left[(\nu^\circ)^2/g\right]^{1/3} \approx 1 \text{ mm}. \tag{6.19}$$

As a consequence, for a *very thin fluid layer*—the *film problem*—when the thickness $d$ of the fluid layer *is of the order of mm, when $Fr_d = O(1)$*, and when the *buoyancy term* in the above leading-order approximate thermal convection equations (6.18a) *is negligible*, instead of the above O-B equations (6.18a), we derive the *classical incompressible model equations* for $v_{i0}$ and $\pi_1$, and the perturbation of the *temperature* $\theta_0$ *is decoupled from this incompressible model.* But, in such a case ($Fr_d = O(1)$), one must *take into account the effect of the free surface tension*, and take into account the so-called *Marangoni effect.*

*For a deeper discussion of this case and the role of buoyancy in the so-called Bénard-Marangoni problem, when the liquid is bounded by a free surface with a temperature-dependent surface tension (thermocapillary instability problem), see our 2009 book [3, Chap. 7], cited in Chap. 1, and also the CISM Course, published with M.G. Velarde, Udine, 10–14 July 2000.*

The above O-B model equations (6.18a) are only valid if the Boussinesq number *Bo* defined by (6.16b) *is bounded* [of the order of O(1)], because *only for this case can we neglect* the above two terms:

$$\alpha \mathrm{Bo}(X_3 - 1)D\theta/D\tau \text{ and } (\mathrm{Bo}Fr_d^2/2)\left[\partial v_i/\partial X_j + \partial v_j/\partial X_i\right]^2,$$

in *the dominant equation for* $\theta$. As a consequence, in the above shallow thermal convection equations (6.18a), the equation for $\theta_0$ is a consistent leading-order approximate model only if

$$\mathrm{Bo} = O(1) \Rightarrow d \approx C^\circ \Delta T/g. \tag{6.20}$$

But if $\mathrm{Bo} \gg 1$, such that $\mathrm{Bo}\, Fr_d^2 = \alpha\, [\mathrm{Bo/Gr}] = O(1)$, then in this case, we obtain the following so-called deep-convection equation for $\theta_0$:

$$\begin{aligned}[1 + \delta(1 - X_3)]D^\circ\theta_0/D\tau = {} & (1/\mathrm{Pr})\Delta_3\theta_0 \\ & + (\delta/2\mathrm{Gr})\left[\partial v_i/\partial X_j + \partial v_j/\partial X_i\right]^2,\end{aligned} \tag{6.21a}$$

with a new parameter:

$$\delta = \alpha \mathrm{Bo} = \ni g d/C^\circ, \tag{6.21b}$$

and for the deep thickness of the liquid layer *d*, we obtain the relation

$$d \approx C^\circ/\ni g. \tag{6.21c}$$

Concerning the *mathematical analysis* of the Bénard problem for deep convection, the reader can find various rigorous results in the papers by *Charki (1996)*. For the *usual liquids*, according to (6.21c), the thickness of the liquid layer should be *at most of the order of the kilometer*!

Naturally, for *technological applications* of the O-B equations, the estimate (6.21c) is very good, but for various geophysical applications, the deep convection equation (6.21a) for $\theta_0$, which takes into account *viscous dissipation,* is more convenient, and the reader can find some numerical calculations in Errafiy and Zeytounian (1991).

It is interesting to note that the *parameter* Bo $(Fr_d)^2$ plays the role of a reference squared Mach number, such that

$$(M_{conv})^2 = (\nu^\circ/d)^2/(C^\circ \Delta T), \tag{6.22}$$

based on the temperature difference ΔT between the rigid lower and upper boundaries, with $C^\circ \Delta T$ playing the role of a characteristic acoustic speed for the weakly expansible liquid. Since $Bo = O(1)$, for the rigorous RAM derivation of the model O-B equations (6.18a), the asymptotic expansions (6.17a) are indeed expansions relative to $(M_{conv})^2$, and as a consequence *the derivation of these O-B model equations* from the full NSF convection equations *is really related to the low-Mach number (hyposonic) flow problem* [2], cited in Chap. 1.

We note also that, if Δp is the characteristic pressure fluctuation, then we have $\Delta p/\Delta T \ll C^\circ \rho^\circ$, and Δp is always smaller than the ΔT. This property also justifies to a certain extent the specifying equation (6.15a), viz., $\rho = \rho(T)$, which was adopted at the beginning.

Concerning (6.15a), to be precise, one should note that, from the relation (6.16d) for π (proportional to $(Fr_d)^2$), to leading order, according to expansions (6.17a), it is clear that the presence of the pressure p [in π, by (6.16d)] in a full baroclinic equation of state $\rho = \rho(p,T)$ instead of (6.15a), *does not change the form of the O-B leading-order model equations* (6.18a) in the limit (6.17b) with (6.17c).

Curiously (and erroneously), in the paper by *Rajagopal et al. (1996)*, the authors assert that: "... consequently, [the derivation of the O-B equations] is free from the additional assumptions usually added on in earlier works in order to obtain the correct (O-B) equations." But it is indeed true that the above O-B model equations are *singular in the vicinity of the initial time* $\tau = 0$, where the initial conditions, with given data, are imposed (for **u**, ρ, and T), to initiate the starting NSF dominant dimensionless convection equations (see, for example, pp. 49–52 and pp. 363–369 in [25], cited in "Introduction"). *A detailed analysis in the vicinity of* $\tau = 0$ *is necessary, and this leads to a problem of adjustment to the Boussinesq state*!

Finally, for the functions

$$\begin{aligned} U_i &= v_{i0},\ i = 1, 2, 3,\ \Theta = X_3 - 1 + \theta_0, \\ \Pi &= Gr\, X_3[(X_3/2) - 1] + \pi_1, \end{aligned} \tag{6.23}$$

instead of the O-B model problem (6.18a) with (6.18b), we obtain the classical Rayleigh-Bénard rigid-rigid shallow convection problem (which is the only case corresponding to Bénard's 1900 experiments), as in Drazin and Reid (1981).

In the paper by *Bois (1991)*, the reader will find a pertinent review of studies devoted to asymptotic aspects of the Boussinesq approximation for gases and liquids, and on p. 76 of our book [25], cited in "Introduction", the reader will find a particularly edifying conclusion concerning the RAM approach. On pp. 73–76 of [25], cited in "Introduction", the curious reader will find the *second-order model equations* following from Hills and Roberts' convection equations (expounded on pp. 52–54 of [25], cited in "Introduction"), *first thanks to our RAM approach* [the system of equations (3.34a)–(3.34c)], *and then the system of equations derived in an ad hoc manner by Hills and Roberts in 1991*, where some first-order

**The R-B shallow thermal convection**
for a not very thin liquid layer

$Fr^2 << 1 \Rightarrow d >> [\nu^2/g]^{1/3} \approx 1mm$

$\alpha = \ni \Delta T << 1$ with $\alpha/Fr^2 = Gr = O(1)$

$0 < \Delta T = [T_1 - T°] \approx (gd/C°)$

the free-surface deformation and viscous dissipation are negligible

$D°\theta_0/D\tau = (1/Pr)\Delta_3\theta_0$

---

**The Deep thermal <à la Zeytounian> convection**

$Bo >> 1$ with $\alpha Bo \approx 1 \Rightarrow d \approx (C°/\ni g)$

$\delta = \ni gd/C° \approx 1$ is the deep parameter

the free-surface deformation is negligible but the viscous dissipation is taken into account

$$[1 + \delta(1 - X_3)]D°\theta_0/D\tau =$$

$$(1/Pr)\,\Delta_3\theta_0 + (\delta/2Gr)\,[\partial v_i/\partial X_j + \partial v_j/\partial X_i]^2$$

---

**Bénard- Marangoni convection**

$Fr^2 \approx 1 \Rightarrow d \approx [\nu^2/g]^{1/3} \approx 1mm$

$Gr \approx \alpha$ – no buoyancy and viscous dissipation is negligible

$Bo << 1 \Rightarrow 0 < \Delta T = [T_1 - T°] >> (g\nu)^{2/3}/C°$

$\downarrow \rightarrow \rightarrow \rightarrow \rightarrow d << C°\Delta T/g$

with very complicated upper conditions at the deformable free-surface

**Fig. 6.5** The three main facets of Bénard convection heated from below for a weakly expansible liquid ($\alpha \ll 1$), according to [31], cited in Chap. 1

compressibility effects are taken into account with various poorly balanced terms (see the system of equations p. 76 of [25], cited in "Introduction") (Fig. 6.5).

The above example relating to Hills and Roberts' 1991 paper clearly demonstrates that, *even if an ad hoc derivation is often able to give a valid working model to leading order, in spite of the fact that a deficient approach may have been chosen, such an approach will never be able to derive in a consistent way a well-balanced rational second-order approximate working model which can be associated with the derived leading-order working model and which takes into account some of the physical effects neglected in the leading-order model. This key observation is one of the main reasons for the publication of the present book "Challenges in Fluid Dynamics".*

I would like to mention the recent book *(2012) "Falling Liquid Films"*, published (thanks to *much effort by M.G. Velarde)* with *Kalliadis, Ruyer-Quil, and Sheid.* I had no hand in it, in spite of the fact that I had many discussions and various collaborations relating to *wavy dynamics on a liquid film flow down an inclined plate*, with *Velarde, Ruyer-Quil, and Sheid at the Instituto Pluridisciplinar UCM* in Madrid (Spain) during the years 2000–2004, and published a paper entitled *Thermocapillary long-waves in a liquid film flow* with the four above-mentioned authors in *J. Fluid Mechanics* ***558***, 2005, pp. 199–222 and pp. 223–244. My *name* is completely ignored in this book, except in the references of our book *(CISM Courses and Lectures, N° 428, Udine, 2002),* edited with Velarde, where it is mentioned on p. 434. It stands to reason that my scientific participation, through my visits to Madrid, is obvious, and I do not understand the motive for such poor conduct! Maybe the publication of my 2009 book entitled "*Convection in Fluids*" is the cause and effect of this unusual behaviour. *However, this undoutedly regrettable oversight does not affect my recollection of the years with Velarde in Madrid!*

Before going any further, in the above-mentioned 2012 book, the reader may find Appendix A useful. Here the authors give hommage to *two key Scientists, P.L. Kapitza* and *C.G.M. Marangoni*, who made *pioneering contributions* to the study of falling liquid films and surface-tension-gradient phenomena. Kapitza is already well known as an *outstanding physicist*, for winning the *Nobel Prize for Physics* in *1978*, and also for his *involvement* in the *history of the Soviet atomic project, his disagreement* with *Beria*, and *several critical letters* to *Stalin.*

Let me quote here from the *preface* of the 2002 *CISM Courses and Lectures*:

> As the waves are long in the wavy dynamic on a liquid film flow down an inclined plate, compared to the very thin film thickness, or equivalently, deformations of the free surface are weak, the viscosity of the fluid ensures a great coherence of the flow across the film. These fortuitous characteristics inherent to the falling film problem enable us to drastically reduce the complexity of the basic equations and to obtain systems of simplified model equations.
>
> The advantage of these models is to isolate the underlying physical mechanisms of the phenomena associated with the nonlinear wave evolution on a falling film and to simulate them extensively at a reduced analytical and numerical cost.

In "*Convection in Fluids*", 2009 [25], cited in "Introduction", the reader can find the application of our RAM approach to the derivation of many such *reduced long-wave evolution equations*: *Benney, KS, KS-KdV, IBL, and so on.*

# Chapter 7
# Some Concluding Remarks About Part II

## 7.1 Low Reynolds Number Case

First, concerning the low Reynolds number fluid flows $\mathrm{Re} \ll 1$, the reader will find in Chap. 9 of our 2002 book FMIA 64 [7], cited in "Introduction", various asymptotic modelling aspects of unsteady low Reynolds number flows, which are mainly the results of *Bentwich and Miloh* during the years 1978–1985. In fact, the solution obtained by these authors represents the *entire process of transition from stagnation to the steady state* envisaged by *Proudman and Pearson (1957)*. However, in the paper by *Sano, 1981* <1>, it is shown that the matching procedure proposed by *Bentwich and Miloh (1978)* is incomplete!

As a consequence, in the case of $\mathrm{Re} \ll 1$, the unsteady problem is more complicated and has *a three-region structure*:

(1) a small-time domain,
(2) a large-time inner domain,
and
(3) a large-time outer domain.

This is shown schematically in Fig. 7.1, reproduced from p. 346 (Fig. 9.2) of the 2002 book [7], cited in "Introduction".

## 7.2 Low Mach Number Case

The reader can also find models for the far field in the case of low Mach number flow $\mathrm{M} \ll 1$ on pp. 178 to 184 of our book, 2002 [13], cited in "Introduction", with the various regions sketched in *Fig*. 8.1 at the bottom of p. 183, reproduced here in *Fig*. 7.2. According to this figure, we have in fact a *three-region singular problem*, with regions (I), (II), and (III) and two *matching* regions (V) and (VI), and the


R.Kh. Zeytounian, *Challenges in Fluid Dynamics*,
DOI 10.1007/978-3-319-31619-2_7

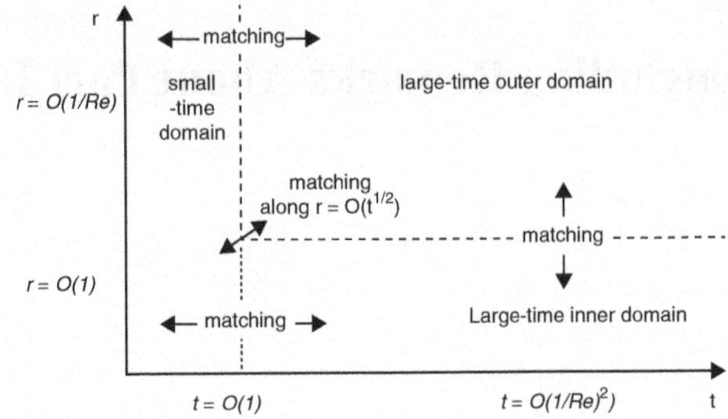


**Fig. 7.1** Schematic sketch demonstrating the matching procedure in the case of unsteady low ($Re \ll 1$) Reynolds asymptotics

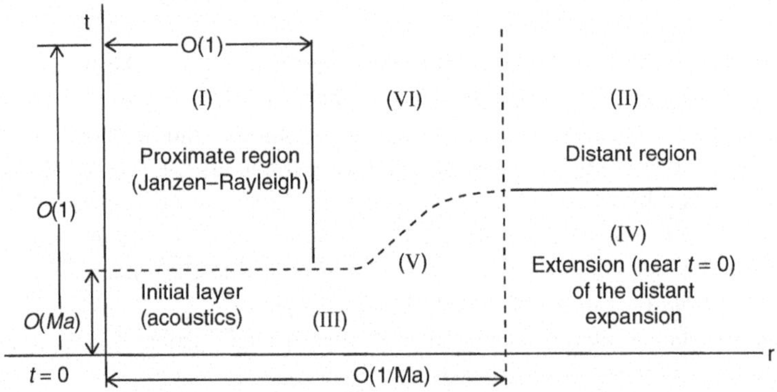


**Fig. 7.2** Sketch of the various regions related to low Mach number limit $Ma \to 0$

extension region (IV) of the *distant acoustic region (II)* (*near* the initial time, for *small times*).

The rather laborious asymptotic theory for the far field was performed by Guiraud (unpublished notes); see also the thesis *by Sery-Baye (1994)*. In the latter, the reader can find the various results, which extend those of *Viviand (1970)* and also those of *Leppington and Levine (1987)*.

## 7.3 Triple-Deck Model

Concerning the singular coupling and the triple-deck model, the reader will find in the last paper by Guiraud [7], cited in Chap. 1, entitled "*Going on with Asymptotics*", the *early roots* of such a theory, but also the *upstream influence mystery*, *Lighthill's theory*, the main asymptotic features, and a *few examples* of *significant achievements*.

This singular coupling and the triple-deck model are expounded in Chap. 12 of our book FMIA 64, *2002* [7], cited in "Introduction" (see pp. 471 to 525 and references on pp. 542 to 545). Here I quote a few lines extracted from a recent paper by Germain [1], cited in the references of the Preface:

> But, again, I express the view that, even a genius would not have been able to build the whole of the triple-deck model without the help of the matched asymptotic expansion techniques. Triple-deck theory is now a very important building block in the new fluid dynamics inspired by asymptotics, and it may be fully included within the heritage of Prandtl.

A good review of different contributions issued from this theory may be found in *Stewarston* [23], cited in "Introduction", *Smith (1982)*, and *Sychev (1972)*, who gave a quite satisfactory answer to Prandtl's query about separation. But the fact still remains that, the above-mentioned paper by Guiraud [7], cited in Chap. 1, is for me more illuminating when we come to consider the triple-deck theory!

Here I will say only that our understanding of viscous incompressible (2D steady) flow past a finite plate, at Re $\gg$ 1, has been revolutionized by application of the triple-deck theory to the flow near the trailing edge (see Fig. 7.3, which is Fig. 4.1 on p. 92 of [13], cited in "Introduction"). The small parameter is $\varepsilon = 1/\mathrm{Re}$. It comprises a region around the trailing edge whose extent is of order $\mathrm{Re}^{-3/8}$ times the length of the plate, and which matches upstream with the classical *Blasius solution* and its associated *outer flow,* and *downstream* with the *two-layered wake* analyzed by *Godstein (1930)*. There is also a *small circular core,* and just as in the corresponding vicinity of the leading edge, the full incompressible Navier equations apply.

The *viscous–inviscid interactive* problem is also relevant to *application* of the *triple-deck theory*, and there is a recent paper by *Veldman (2001)*, which presents his personal view of the history of *viscous–inviscid interactive methods. According to Veldman,* the *subdivision of the flow field* around an *airfoil in an inviscid flow region and in a viscous shear layer is sketched (with exaggerated thickness) in Fig. 7.4 (reproduced from p. 263 of* [14], cited in "Introduction"). It is now well known that the main challenge in solving Prandtl's boundary-layer equations has been to overcome the singularity at a point of steady-state flow separation and that the Stewartson-Neiland-Sychev triple-deck theory has inspired a solution to this challenge. The interested reader will find more information in pp. 263 to 266 of [14], cited in "Introduction".

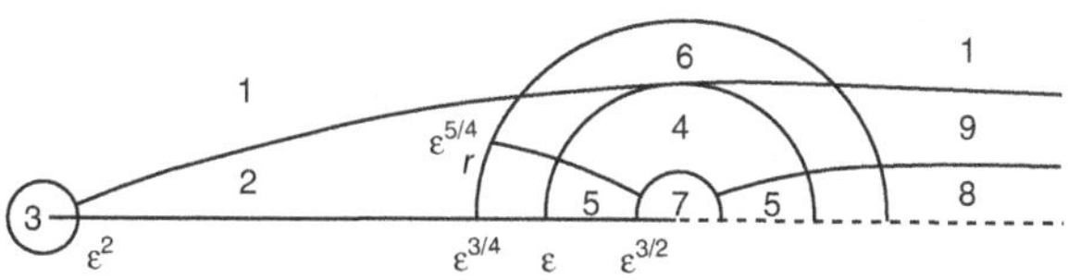


**Fig. 7.3** (1) Potential flow, (2) Blasius BL, (3 and 7) Navier incompressible region, (4) Main deck, inviscid, (5) Lower deck, viscous, (6) Upper deck, potential flow, (8) Inner Golstein wake, and (9) Outer Goldstein wake

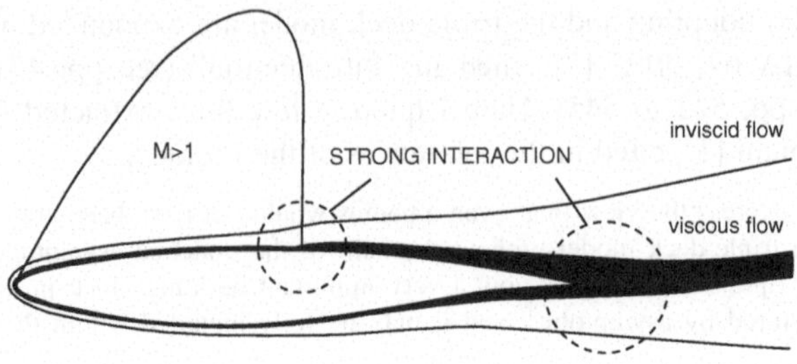


Fig. 7.4 Subdivision of the flow field around an airfoil

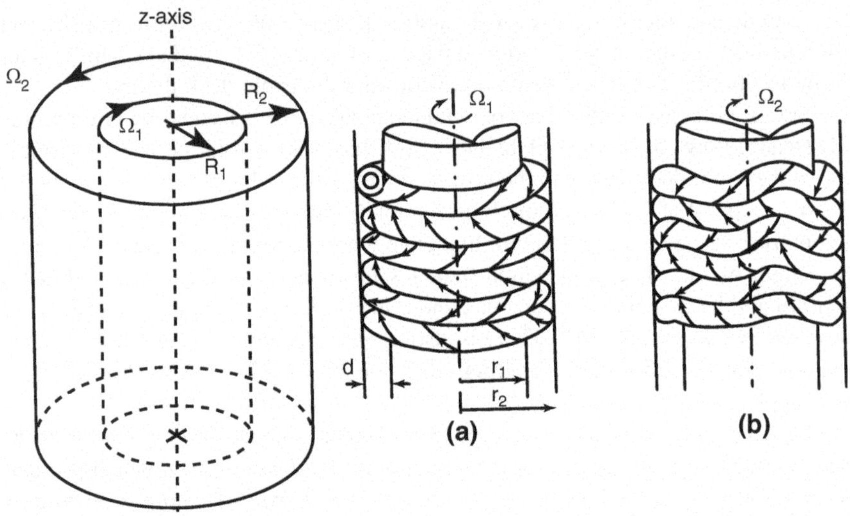


Fig. 7.5 The Couette-Taylor problem

## 7.4 Couette-Taylor Problem

A very classical problem in fluid dynamics is the famous *Couette-Taylor* problem (a French-English one!), concerning the flow between two (infinite) concentric circular cylinders in the framework of linear theory. Here, we shall give only a very *short account* of *more recent results* concerning the nonlinear (in fact, in *weakly nonlinear*) theory, which has been thoroughly investigated, in particular thanks to the theoretical results of *G. Iooss* [see the b*ooks by Iooss and Adelmeyer (1992), and Chossat and Iooss (1994)*], using *center manifold reduction, bifurcations*, and *amplitude equations*.

In particular, in Chap. II of the book by *Iooss and Adelmeyer (1992)*, the reader will find *applications of this center manifold approach, normal forms, and bifurcations of vector fields near critical points* to the Couette-Taylor problem, which is a *paragdim* for the *transition* from *regular to turbulent behavior and chaos*, discussed in Chap. 10 (pp. 387 to 448) *of our book "Theory and Applications of Viscous Fluid Flows" (2004). Figure* 7.5 *shows a view of the Couette-Taylor*

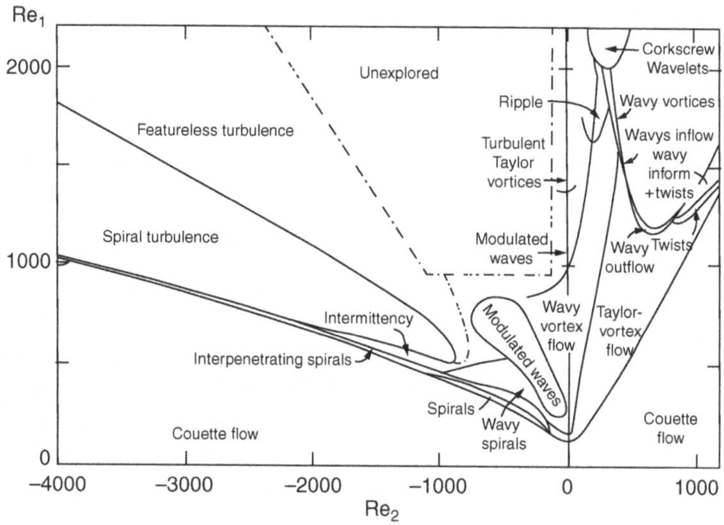


**Fig. 7.6** The Couette-Taylor problem: experimental stability diagram

*problem taken from Fig. 9.8 (on p. 370) and Fig. 9.9 (on p. 372) of [14]*, cited in "Introduction": case (a) *Taylor vortices*, and case (b) *wavy vortices and azimuthal* waves.

*Figure* 7.6, *which is Fig. 9.10 on p. 373 of* [14], cited in "Introduction", *shows an experimental stability diagram for the Couette-Taylor problem, where* $\mathrm{Re}_1$ *and* $\mathrm{Re}_2$ are the Reynolds numbers of the inner and outer cylinder (see Anderek et al. 1986). This *stability diagram is interesting, but* I am not so sure that it is possible to give a theoretical justification for a such large family of cases!

## 7.5 Meteo–Fluid Dynamics

Finally, in Sects. 8.6 and 8.7 (in *Part III*), two meteo problems are considered, namely, *lee waves* and *breeze phenomena*, respectively *downstream of a mountain* and as a *consequence of a local thermal effect between the land and the sea. Working models are derived in both cases. Figure* 7.7, *which is taken from p. 169 of* [13], cited in "Introduction", shows a typical result of the 3D calculation of wet zones above the Bassin d'Arcachon, in France, for a stable stratification. In pp. 168–170 of our book (2002) entitled *"Theory and Applications of Non-viscous Fluid Flows"*, the reader will find some details concerning this calculation via a rational working model, and on pp. 29–33 of our companion 2014 book, the linear 3D steady equations are given with various comments.

The numerical calculation by *Trochu (1967), carried out at the Météorologie Nationale, Quai Branly (Paris),* is based on a representation of the coast as an

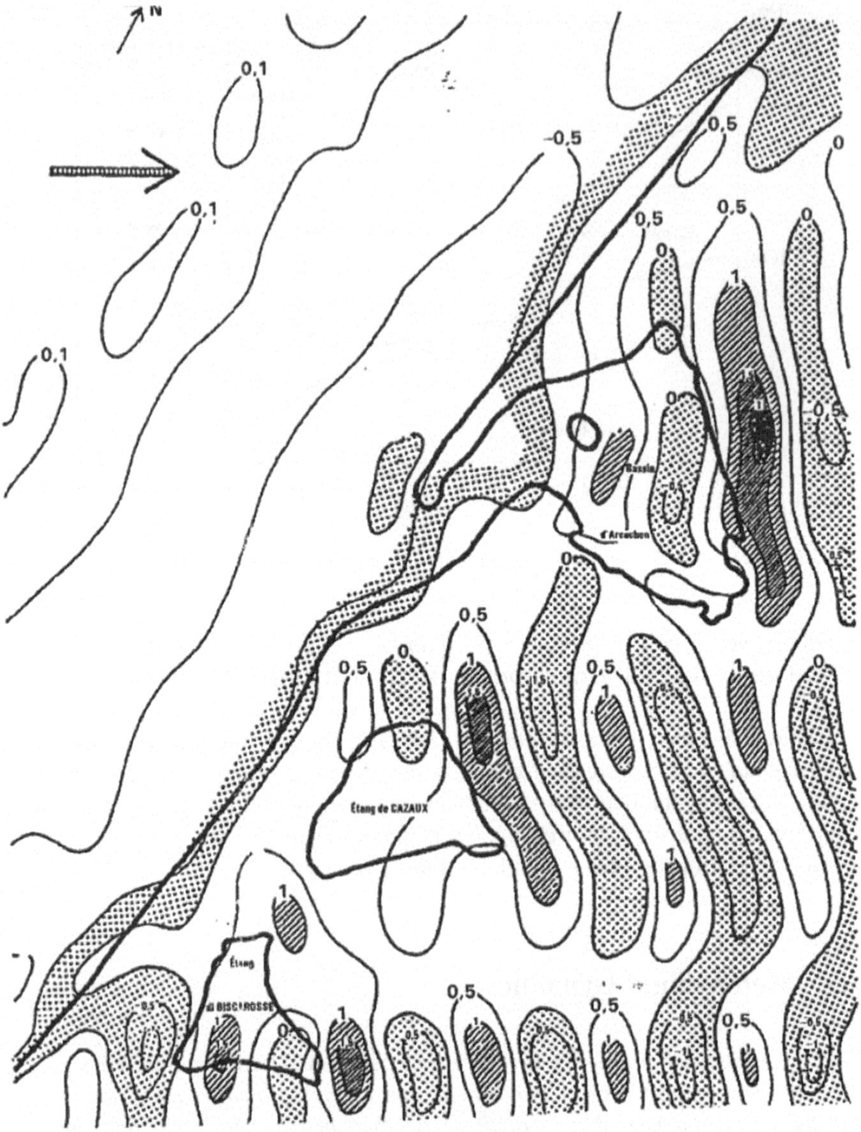


**Fig. 7.7** Wet zones above the Bassin d'Arcachon for stable stratification

algebraic sum of typical simple obstacles in the form of paraboloids of revolution—observing that coast, which separates the sea from the land, plays the role of an obstacle (*mountain*!) due to the difference in the roughness of the land surface relative to that of the sea. With the coast represented by such a sum, the influence of each paraboloid at each point of the region investigated was calculated, and the sum of these influences obtained. The results with Kibel in Moscow were obtained for various altitude levels. In *Fig. 7.7:* (–) negative*: broken line, and* (+) positive*: hatched.*

## 7.6 Some Complementary Remarks

It is clear (at least for me!) that at the present time a gap still exists between the RAM approach and high-speed computer simulation! In many cases the results of numerical computations, which are often fascinating, do not correspond satisfactorily with experimental/laboratory visualizations, mainly because we do not have at our disposal a criterion whereby we can establish the limit of validity of the working models that have been used!

It is rare in scientific publications to encounter the theoretical treatment of a full unsteady Newtonian fluid flow problem directly inspired by technological problems, and it seems to me that there is a lack of *theoretically consistent, valid, non-ad hoc* studies which give at least a simple description of such fluid flows via model problems that can be used to explain them using numerical simulation. *While* the *increasing capacity of supercomputers makes it possible to perform very valuable computations,* it *seems to me obvious that only the use of a valid reduced rational working model emerging from the fluid dynamics equations with consistent initial and boundary conditions can provide a way to obtain realistic numerical results! On the contrary, any* reference to the nature of the selected working model—often ad hoc, or deduced rationally—and its relationship with the real stiff starting problem is *systematically omitted*!

*Indeed, information about the process of the fluid dynamics modelling in numerical simulation/computation is often simply completely overlooked, or treated as a "futile" question! The crucial problem of initial conditions (with given data), which strongly influence the subsequent formation and evolution of fluid flows, is also usually overlooked, despite the fact that it is very often necessary to consider an unsteady adjustment problem, due to the singular nature of the main derived model (with fixed time) in the initial time region.*

I am convinced that, *the more computing and numerical algorithmic processes become efficient with the use of high-speed computers, the greater will be the need for conceptually consistent techniques capable of unravelling stiff fluid-flow problems. The RAM approach proves to be one of the most efficient tools among these. This (perhaps optimistic) statement was, during the years 1970–1990, a guideline for me, as I began working on the asymptotic modelling of atmospheric flows!*

Concerning applied mathematicians, it seems that the situation is somewhat paradoxical! Inspired by the last published paper by *Germain* [1], cited in the references of the Preface, devoted to the *"New Mechanics of Fluids of Ludwig Prandtl"*, I would like to make a final point about attempts to obtain *rigorous 'pure' mathematical results for fluid dynamics:* I recognize the *need* for such *rigorous results* by *mathematicians* as an application of their imaginative approach via *qualitative nonlinear abstract functional analysis,* despite the fact that these results give only a *qualitative idea,* and *do not* in reality *have any practical interest in modelling real stiff problems.*

In the end, it seems to me also of great importance that a *rational approach be adopted to make sure, for example, that at least the terms neglected in* the *initial reduced working equations really are much smaller than those retained* by *numericians* in their approximate working fluid dynamics models for simulation/computation! Until this is done, and it is already possible, thanks in part to the RAM approach, it will be difficult to convince the detached and possibly skeptical reader of their value as an aid to understanding.

I end these mainly critical remarks by referring *to the difficulty I had as a young mathematician, a postgraduate student of Serget Nikitovich Mergelyan, to understand the ad hoc introduction of the two times in the book by Kibel (1957), devoted to weather forecasting, 58 years ago in Moscow, in Il'ya Afanas'evich Kibel's Department of Dynamic Meteorology*. In our 2014 book, which takes the form of a scientific autobiography, Chap. 4 develops the basic NSF equations in a historical context (and this Chap. 4 has inspired our present Chap. 1).

Finally, as was so well observed by Robert E. O'Malley, Jr. in the Books Review of SIAM Review (p. 569, downloaded 10/06/14):

> *Overall, readers will realize the unique value of Zeytounian's work and perspective and will come to appreciate his willingness to tackle very difficult problems, aiming to help fluid dynamical "numericians". An example is modelling turbomachinery using the number of blades on a rotor as a large parameter.*

This is for me a good appreciation, mainly concerning applications of the RAM approach to very stiff fluid dynamics initial–boundary value problems! I hope that the reader will agree with Robert E. O'Malley, Jr after reading the above Parts I and II.

The following part of the book, *Part III, also the last, is a miscellany* of working models derived for various stiff fluid dynamics problems during the last five decades. These are, in fact, expounded in my various books published by Springer during the years 1974 to 2014, which give an overview concerning the possibilities made available by our RAM approach! However, in Sects. 8.2 to 8.9, I have to draw up a rather different version, where the various facets of our RAM approach are most clearly revealed!

I should stress that the *RAM/deconstruction approach to the NSF system* of (exact!) equations allows one to obtain a rather *precise estimate of what has been neglected in each of the main localised reduced and simplified working models for numerical simulation/computation, which are then reconstituted via matching! Such an estimate is correlated with the numerical value of the dimensionless parameters whose value has been neglected in getting each reduced and simplified working model!*

## Reference

<1> Sano T (1981) Unsteady flow past a sphere at low Reynolds number. J Fluid Mech 112:433–441

# Part III
# Miscellaneous: Various Fluid Dynamics Workings Models

# Chapter 8
# Some Applications of the RAM Approach During the Years 1974–2014

## 8.1 Introduction

Sections 8.2–8.9 may be useful as a reference for scientists, and in particular, applied mathematicians and engineers, working in the fields of Newtonian fluid dynamics, especially in the areas of modelling, simulation, and numerical computations of fluid flow phenomena. This chapter should also prove to be of interest to individual applied mathematicians working in the areas of physics and mathematics.

While the relevant problems have been solved and expounded in my various publications during the years 1974–2012, only an outline of the solution is sometimes given, and the reader must supply the details. The author recommends that students seek to obtain their own solutions to these problems, in order to gain familiarity with the techniques. The reader will observe that it is often necessary, when applying the RAM approach to the above problems, to devote much time to an analytical inquiry in order to obtain a useful working model, ready for a numerical computation.

On the other hand, concerning *"asymptotics"*, which is an essential tool in our RAM approach, the reader can find an application below which clearly demonstrates the importance of *matching*! In particular, we consider as starting equations the *steady 2D Navier system* of two equations, written in dimensionless form, viz.,

$$\nabla \cdot \mathbf{u} = 0, \tag{8.1a}$$

$$(\mathbf{u} \cdot \nabla)\mathbf{u} + \nabla p = (1/\mathrm{Re})\{\mu(\mathrm{n})\nabla^2\mathbf{u} + (\mathrm{d}\mu/\mathrm{dn})\,[\partial\mathbf{u}/\partial\mathrm{n} + \nabla \mathrm{v}\}, \tag{8.1b}$$

with

$$\mathbf{u} = 0 \text{ on the body surface } \mathrm{n} = 0, \tag{8.1c}$$

for the velocity $\mathbf{u} = \mathrm{u}\,\boldsymbol{\tau} + \mathrm{v}\,\boldsymbol{\nu}$, and pressure $p$, where


R.Kh. Zeytounian, *Challenges in Fluid Dynamics*,
DOI 10.1007/978-3-319-31619-2_8

$$\nabla = \mathrm{h(s)}\ (\partial/\partial \mathrm{s})\tau + (\partial/\partial \mathrm{n})\mathbf{v}, \tag{8.1d}$$

with $\mathrm{h(s)} = [1 + K(\mathrm{s})\ \mathrm{n}]^{-1}$ the curvature of the body surface, a bounded function such that

$$(1/\mathrm{Re})\ \mathrm{d}K(\mathrm{s})/\mathrm{ds} \to 0 \text{ as } 1/\mathrm{Re} \equiv \varepsilon^2 \to 0. \tag{8.1e}$$

For the *dynamic viscosity coefficient* $\mu(\mathrm{n})$, it is assumed that

$$\mu(\mathrm{n}) = 1 + \mu^*(\mathrm{n}/\Delta(\varepsilon)), \tag{8.1f}$$

with $\mu^*(\infty) \to 0$, and $\Delta(\varepsilon) \ll \varepsilon$.

We observe that, if L° is the reference length associated with the shape of the body, then

$$\Delta(\varepsilon) = l^\circ/\mathrm{L}^\circ \ << \varepsilon << 1, \tag{8.1g}$$

and $l^\circ$ is a characteristic length for the *microscopic variation* of the *dynamic viscosity coefficient*!

We must first consider an "*outer limit*".

### 8.1.1 Outer Euler limit

$$\begin{aligned} &\mathrm{Lim}^{\mathrm{E}}{}_{\varepsilon\to 0} \equiv [s \textit{ and } n \text{ are fixed}],\\ &\mathrm{Lim}^{\mathrm{E}}{}_{\varepsilon\to 0}(\mathbf{u},\ \mathrm{p}) = (\mathrm{u}_0,\ \mathrm{v}_0,\ \mathrm{p}_0), \text{for incompressible}\\ &\text{Euler equations, and next order, } (\mathrm{u}_1,\ \mathrm{v}_1,\ \mathrm{p}_1), \text{for linearized Euler}\\ &\text{perturbation equations relative to the preceding Euler equations.}\end{aligned} \tag{8.2a}$$

The following asymptotic expansion is associated with the outer Euler limit:

$$V^{\mathrm{E}} = (\mathrm{u},\ \mathrm{v},\ \mathrm{p}) = V_0{}^{\mathrm{E}} + \varepsilon V_1{}^{\mathrm{E}} + \ldots \tag{8.2b}$$

Then, we have the following intermediate limit.

### 8.1.2 Intermediate Prandtl limit

In this case, instead of (8.2b), we write

$$\begin{aligned} u &= U_0 + \varepsilon U_1 + \ldots, \\ v &= \varepsilon V_1 + \varepsilon^2 V_2 + \ldots, \\ p &= P_0 + \varepsilon P_1 + \ldots, \end{aligned} \tag{8.3a}$$

with

$\mathrm{Lim}^{Pr}{}_{\varepsilon\to 0} \equiv [s,\ and\ N = n/\varepsilon$ are fixed$]$,
$\mathrm{Lim}^{Pr}{}_{\varepsilon\to 0}(\mathbf{u},\ p) = (U_0,\ V_1,\ P_0)$, for the Prandtl classical incompressible BL equations, and second-*order*, $(U_1,\ V_2,\ P_1)$. (8.3b)

The BL equations for $(U_0, V_1, P_0)$ are

$$\begin{aligned} &\partial U_0/\partial s + \partial V_1/\partial N = 0, \\ &U_0 \partial U_0/\partial s + V_1 \partial U_0/\partial N + \partial P_0/\partial s = \partial^2 U_0/\partial N^2, \\ &\partial P_0/\partial N = O, \end{aligned} \tag{8.4a}$$

and using (8.3a) and (8.3b), the following relations are deduced:

$$\begin{aligned} &v_0 = 0 \text{ as } n = 0, \\ &U_0 \to u_0(s,\ 0) = u_{e0}(s) \text{ as } N \to \infty, \\ &P_0 = p_0(s,\ 0) = p_{e0}(s), \\ &dp_{e0}(s)/ds + u_{e0}(s) du_{e0}(s)/ds = 0, \\ &v_1(s,\ 0) = \int_\Delta [du_{e0}(s)/ds - \partial U_0/\partial s]\, dN, \end{aligned} \tag{8.3c}$$

where $\Delta = [0, \infty]$, and after *matching*,

$$\mathrm{Lim}^{Pr}{}_{\varepsilon\to 0} \text{ with } \mathrm{Lim}^{E}{}_{\varepsilon\to 0}, \tag{8.5}$$

We observe that the first relation, $v_0 = 0$ as $n = 0$, in (8.5), is directly related to the Euler non-viscous equations and *leads to the possibility of writing the steady slip condition on the wall n* = *0*, for the usual *steady-state Euler equations.*

For $(U_1, V_2, P_1)$, we obtain the *second-order approximation* to the BL equations, where curvature effects are taken into account! This gives

$$\begin{aligned} &\partial U_1/\partial s + \partial V_2/\partial N = -K(s)\partial(NV_1)/\partial N, \\ &\partial(U_0U_1)/\partial s + V_2\partial U_0/\partial N + V_1\partial U_1/\partial N + \partial P_1/\partial s \\ &= \partial^2 U_1/\partial N^2 + K(s)\{NU_0\partial U_0/\partial s - U_0V_1 + Ndp_{e0}(s)/ds + \partial U_0/\partial N\}, \end{aligned} \tag{8.4b}$$

and matching leads to

$$U_1 \to u_1(s,\ 0) = u_{e1}(s) \text{ as } N \to \infty, \tag{8.3d}$$

$$P_1 = p_1(s,\ 0) + K(s)\int_{\Delta^*} |U_0|^2 dN\,, \tag{8.3e}$$

where $\Delta^* = [N, \infty]$.

But *for the time being*, we *are unable to take into account* the *influence of the microscopic coefficient* $\mu^*(n/\Delta(\varepsilon))$, and for that we envisage a tentative third limiting process, concerning the lower (thin) viscous layer (LVL) very close to body surface!

### 8.1.3 Lower Viscous Limit

$$\mathrm{Lim}^{Lv}{}_{\varepsilon\to 0} \equiv [s \text{ and } \eta = \eta/\Delta(\varepsilon) \text{ are fixed}], \tag{8.6}$$

with a local asymptotic expansion

$$\begin{aligned} u &= \varepsilon^{\alpha} U^*{}_0(s, \eta) + \ldots, \\ v &= \varepsilon^{\beta} V^*{}_0(s, \eta) + \ldots, \\ p &= P^*{}_0(s, \eta) + \varepsilon P^*{}_1(s, \eta) + \ldots . \end{aligned} \tag{8.7a}$$

This limit is envisaged in the hope that we can take into account the influence of the microscopic coefficient $\mu^*(n/\Delta(\varepsilon))$. After some testing,

$$\Delta(\varepsilon) = \varepsilon^{\lambda},\ \lambda > 1 \text{ and then } \lambda = \beta - \alpha, \tag{8.7b}$$

we obtain for ($U^*{}_0$, $V^*{}_0$, $P^*{}_0$) the following reduced system of equations in the thin *lower viscous layer (LVL) model equations*:

$$\partial U^*{}_0/\partial s + \partial V^*{}_0/\partial \eta = 0, \tag{8.7c}$$

$$\partial/\partial\eta\{[1 + \mu(\eta)]\partial U^*{}_0/\partial\eta\} = 0, \tag{8.7d}$$

$$P^*{}_0 = p_{e0}(s), \tag{8.7e}$$

with the conditions (according to no-slip condition (8.1c) for the starting Navier viscous equation):

$$U^*{}_0 = 0 \text{ and } V^*{}_0 = 0, \text{ on } \eta = 0. \tag{8.7f}$$

Now, the last critical step is to determine the "*interaction*" between the BL and the LVL.

### 8.1.4 Interaction Between the BL and the LVL

The solution of (8.7c)–(8.7e) with (8.7f) is easy:

$$U^*_0 = A^\circ(s)\int_{\Delta^{**}} [1 + \mu(\tau)]^{-1} d\tau, \tag{8.8a}$$

where $\Delta^{**} = [0, \eta\,]$, and when $\eta \to \infty$, we obtain

$$U^*_0(s, \eta) \approx A^\circ(s)\,\{\eta + I^\circ + \ldots\}, \tag{8.8b}$$

with

$$I^\circ = \int_\Delta \left\{-1 + [1 + \mu(\eta)]^{-1}\right\} d\eta. \tag{8.8c}$$

As a consequence, *a constraint is imposed, via* $I^\circ$, on the function $\mu(\eta)$*!*

Finally, we take into account the following behavior in the BL for the *tangential component of the velocity* u:

$$u = U_0(s,\ 0) + N\,(\partial U_0/\partial N)_{N=0} + \varepsilon U_1(s,\ 0) + \ldots,\ \text{as } N \to 0, \tag{8.9a}$$

and also the relation

$$N = \varepsilon^{\lambda-1}\eta \tag{8.9b}$$

along with (8.7a). We verify that the matching between the local asymptotic expansion (8.7a) and the asymptotic expansion (8.3a), in the intermediate Prandtl BL, is only possible if

$$\alpha = 1, \lambda = \alpha + 1 = 2,\ \text{and}\ \beta = 3, \tag{8.9c}$$

and in a such case we derive the following relations:

$$\begin{aligned} &U_0(s,\ 0) = 0,\ U_1(s,\ 0) = A^\circ(s)\, I^\circ, \\ &V_1(s,\ 0) = V_2(s,\ 0) = 0, \\ &A^\circ(s) = (\partial U_0/\partial N)_{N=0.} \end{aligned} \tag{8.10a}$$

Indeed, the leading order LVL is active only at the level of the *second-order BL equations* (8.4b), and at $N = 0$ we have the boundary conditions

$$U_1(s,\ 0) = I^\circ(\partial U_0/\partial N)_{N=0},\ V_2(s,\ 0) = 0, \tag{8.10b}$$

where $I^\circ$ is given by (8.8c).

As a final result, we find that the *above conditions* (8.10b) *are conditions at* $N = 0$ *for* the *second-order BL equations* (8.4b), for ($U_1$, $V_2$, $P_1$). From (8.9c), $\Delta(\varepsilon) = \varepsilon^2 = 1/\mathrm{Re}$.

Concerning the problem (8.4b) with (8.10b) for ($U_1$, $V_2$, $P_1$), we don't know whether it is well-posed or not! As a particular case, the classical *Blasius problem* was considered by *Godts* (see his paper 1991 <1> with *Zeytounian*). It seems that the uniqueness of a solution of the problem (8.4b) with (8.10b) for ($U_1$, $V_2$, $P_1$) is guaranteed if we assume that the latter also satisfies the following behaviour at infinity:

$$\lim_{N\to\infty} \partial U_1/\partial N = \lim_{N\to\infty} \partial u_1/\partial n, \text{ where } N = n/\varepsilon, \tag{8.10c}$$

which follows from the matching between (8.2b) and (8.3a) for the vertical derivative $\partial u/\partial n$.

## 8.2 Turbomachinery Homogenized Flows

One cannot escape two facts concerning turbomachinery fluid flows! First, *the blades are usually very closely spaced*, and *second, were the blades infinitely close to each other, the flow would be guided along some manifold, unchanged by any rotation around the axis of the turbomachine.*

Below, we assume that the starting model of Eulerian non-viscous and incompressible fluid flow is adequate. The first such attempt to apply our approach appears in *Guiraud and Zeytounian* [19] and [20], cited in "Introduction", and below the reader will find an asymptotic model relying on a double-scale analysis of the problem of flow within a row of blades, either fixed or rotating.

The sketch in Fig. 8.1 *shows the geometry of an axial compressor with a large number of rows of thickness l*. If $N \gg 1$ is the number of blades per row, we have a small parameter

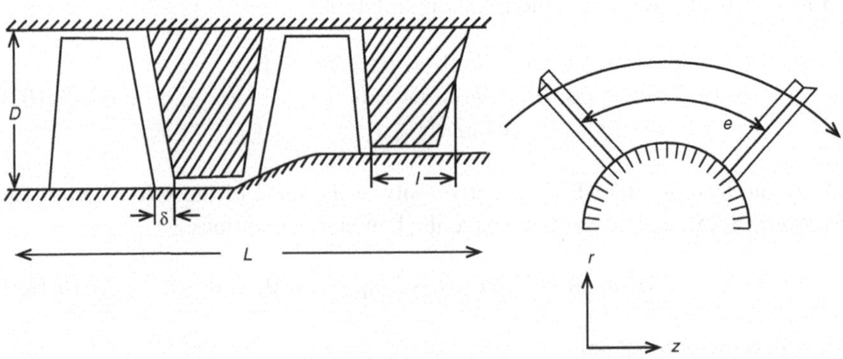


**Fig. 8.1** Sketch of the geometry of an axial compressor

$$\varepsilon = 1/N << 1, \tag{8.11a}$$

which controls the flow and provides a way to apply asymptotics. We consider only the case of the flow within a row, the so-called *trough-flow model* (see Fig 8.1, *which is taken from p. 188 of* [13], cited in "Introduction") and also the *local flow* at *the inlet* and *the outlet* of a row, because the *classical trough-flow in an axial turbomachine (compressor) is invalid near the locus of the leading/trailing edges of a row.*

We use cylindrical coordinates r, θ, z, throughout, with relative velocity components u, v, w, and pressure p. The non-viscous and incompressible fluid flow equations are written below in dimensionless matrix form as

$$\partial \underline{T}/\partial t + \partial \underline{R}/\partial r + \partial \underline{Z}/\partial z + (1/r)\partial \underline{S}/\partial \theta + \underline{H}/r = 0. \tag{8.11b}$$

As the situation considered corresponds to "*long blades*" compared with the breadth of the channel from blade to blade—*via the asymptotics*—the obvious working idea is that *the flow is very close to axisymmetrical when the number of blades N is very large!*

### 8.2.1 A New Formulation

First, it seems necessary to change the usual coordinates (r, θ, z) to a new system of coordinates (r, $\chi$, z) better suited to a row, with the idea that, in the limit (when $\varepsilon \to 0$), the *through flow* will be independent of $\chi$, as shown below in the hierarchy (8.12a). We thus introduce the following change:

$$\theta = \Theta(t, r, z) + 2\pi\varepsilon(k + \chi), \tag{8.11c}$$

where the surface $\theta = \Theta(t, r, z)$ corresponds to the blade skeletons in a row, when $\varepsilon \to 0$, and outside are material surfaces that are the extensions of the blade skeleton. Clearly, without any approximation, we see that the flow has to be periodic in $\chi$, which lies between *zero* and *one;* and for the through flow in the row, we consider the following outer expansion index (k running from 1 to N):

$$(u, v, w, p) \equiv U = U_{k,0}(t, r, \chi, z) + \varepsilon U_{k,1} + \dots . \tag{8.11d}$$

However, the above *outer expansion* (8.11d) will *fail near the ends of the row, where inner expansions are needed, at* $z = h(r)$, *the locus of the leading and trailing edges of the row. We thus write the following inner expansion:*

$$U = U^*_{k,0}(t, r, \chi, z^*) + \varepsilon U^*_{k,1} + \dots, \tag{8.11e}$$

where

$$z^* = [z - h(r)]/\varepsilon. \tag{8.11f}$$

In the new coordinate system (r, $\chi$, z), instead of the matrix equation (8.11b), we obtain

$$\partial \underline{\mathbf{G}}_k/\partial\chi + 2\pi\,\varepsilon\,r\,\underline{\mathbf{L}}_k = 0, \tag{8.11g}$$

where the operator

$$\underline{\mathbf{L}}_k = \partial \underline{\mathrm{T}}_k/\partial t + \partial \underline{\mathrm{R}}_k/\partial r + \partial \underline{\mathrm{Z}}_k/\partial z + \mathrm{H}_k/r, \tag{8.11h}$$

and

$$\underline{\mathbf{G}}_k = \begin{pmatrix} \gamma_k u_k - r(\partial\Theta/\partial r)p_k \\ \gamma_k v_k + (1/\lambda)p_k \\ \gamma_k w_k - r(\partial\Theta/\partial z)p_k \\ \gamma_k + \mu r(\partial\Theta/\partial t) \end{pmatrix} \tag{8.11i}$$

with

$$\gamma_k = \lambda v_k - r\,u_k(\partial\Theta/\partial r) - r\,w_k(\partial\Theta/\partial z) - \mu r\,(\partial\Theta/\partial t), \tag{8.11j}$$

$$\lambda = \omega^\circ\,(D/w_\infty) \text{ and } \mu = D/t^\circ w_\infty, \tag{8.11k}$$

where $t^\circ$ is the reference time, $D$ is the diameter of the row, $\omega^\circ$ is the reference value of the angular velocity $\omega$ of the row, and $w_\infty$ is the upstream uniform axial velocity. Then, the dimensionless pressure is $p = (p' - p'_\infty)/(w_\infty)^2\rho^\circ$, with $\rho^\circ =$ constant density of an incompressible fluid, p′ the pressure with dimensions, and $p'_\infty$ the upstream constant pressure.

### 8.2.2 *Deconstruction via Asymptotics*

#### First Stage

Two facts should be stressed at the outset. *First*, if we assume *axially symmetrical flow*, i.e., $\partial/\partial\chi = 0$, we derive the following reduced equation:

$$\underline{\mathbf{L}}_k = 0, \tag{8.12a}$$

and it may be checked that (8.12a) is the matrix form of *axially symmetrical flow*.

*Second*, if we set $\varepsilon = 0$ by *brute force* in (8.11g), we do not get the equations of *axially symmetrical flow*, but rather, the highly degenerate equation

$$\partial \underline{\mathbf{G}}_{\mathrm{k}} / \partial \chi = 0. \tag{8.12b}$$

This may seem somewhat *strange*, but in fact it should be expected! The way in which it was defined ensures that, when $\varepsilon$ is small, variations in the direction $\chi$ are *magnified by* $1/\varepsilon$ *in comparison with variation with t, r, and z!*

*This is a crucial point,* and indeed, substituting the above outer expansion (8.11d) in (8.11g), we get a hierarchy of approximate equations (we write only the first two):

$$\partial \underline{\mathbf{G}}_{\mathrm{k},0} / \partial \chi = 0; \ \partial \underline{\mathbf{G}}_{\mathrm{k},1} / \partial \chi + 2\pi \mathrm{r} \underline{\mathbf{L}}_{\mathrm{k},0} = 0, \tag{8.12c}$$

which consists of equations to be solved in turn.

## Second Stage

As appropriate to our problem, we chose the solution of the first equation of the above hierarchy (8.12c), $\partial \underline{\mathbf{G}}_{\mathrm{k},0} / \partial \chi = 0$, for which $\mathrm{u}_{\mathrm{k},0}$, $\mathrm{v}_{\mathrm{k},0}$, $\mathrm{w}_{\mathrm{k},0}$, $p_{\mathrm{k},0}$ are all independent of $\chi$. At this stage, *we don't know how these functions depend on t, r, and z*. Then, if we use the second equation of (8.12c) to compute $\mathrm{u}_{\mathrm{k},1}$, and so on, we encounter a *compatibility condition arising from periodicity*, which *enforces* $\underline{\mathbf{L}}_{k,0} = 0$, and we thus obtain an *axially symmetrical through-flow,* theory.

In fact, the *interesting* and *unexpected point* is that we may go a step further and get a *through-flow theory to order* $\varepsilon$! For this, we assume that the channel between two consecutives blades is

$$0 \leq \chi_{\mathrm{e}} \leq \chi \leq \chi_{\mathrm{i}} \leq 1, \text{ where } \Delta(\mathrm{r}, \mathrm{z}) = \chi_{\mathrm{i}} - \chi_{\mathrm{e}},$$

and we define an *average of U* (denoted by <*U*>) and also the *bracket of U* (denoted by [*U*]) as the *jump* in *U* from blade to blade:

$$< U >= (1/\Delta) \cdot \int_{\chi_i}^{\chi_e} U \mathrm{d}\chi, \text{ and } \ [U] = U_{\mathrm{i}} - U_{\mathrm{e}}. \tag{8.12d}$$

In particular, the *bracket for the pressure* may be viewed as the *pressure difference between the two sides of one and the same blade*!

### 8.2.3 *The Guiraud-Zeytounian Working Model (G-Z 1971)*

Up to first order in ε, the *averaged/homogenized G-Z 1971 approximate reduced through-flow working model is the following.* The average velocity <**v**> and pressure <*p*>, viz.,

$$< \mathbf{v} >= \mathbf{v}_{k,0} + \varepsilon < \mathbf{v}_{k,1} > \text{, and } < p >= p_{k,0} + \varepsilon < p_{k,1} > , \tag{8.13a}$$

satisfy, with an *error of* $\varepsilon^2$, the *axially symmetrical through-flow equations*

$$\begin{aligned} &\partial\Delta/\partial t + \mathrm{div}(\Delta < \mathbf{v} >) = 0(\varepsilon^2), \\ &\partial < \mathbf{v} > /\partial t + [\mathrm{Curl} < \mathbf{v} > + 2\Omega \mathbf{e}_z] \wedge < \mathbf{v} > + \mathrm{grad} J \\ &- (1/\Delta)\Pi \mathrm{grad}\,\Sigma = 0(\varepsilon^2), \end{aligned} \tag{8.13b, c}$$

where

$$J = < p > + (1/2)| < \mathbf{v} > |^2 - (1/2)\Omega^2 r^2, \; \Omega = \lambda\,\omega. \tag{8.13d}$$

### 8.2.4 *Some Final Remarks*

Two points should be stressed. *First*, the breadth Δ, of the channel from blade to blade enters into the continuity equation (8.13b) in an obvious way, and *second, there is a source term in the momentum equation* (8.13c) for <**v**>, namely,

$$\boldsymbol{F} = (1/\Delta)\Pi \mathrm{grad}\,\Sigma, \tag{8.14a}$$

which is proportional to the jump in pressure:

$$\Pi = (1/2\pi)\{[p_{k,1}] + \varepsilon\,[p_{k,2}]\}, \tag{8.14b}$$

and orthogonal to

$$\Sigma = S + 2\pi\varepsilon\{(1/2)\,(\chi_i + \chi_e)\} = \text{const.}, \tag{8.14c}$$

where $S = \Theta - \theta$ and Σ is a material surface which is just in the middle of the channel, from blade to blade:

$$\partial\Sigma/\partial t + < \mathbf{v} > \mathrm{grad}\,\Sigma = 0(\varepsilon^2). \tag{8.14d}$$

The force ***F*** is such that

$$\boldsymbol{F}.\mathrm{Curl}\,\boldsymbol{F} = 0, \tag{8.14e}$$

and it has long been known in classical through-flow theory. It arises *from redistribution/homogenization of forces exerted on the flow by the blades.*

*This above derivation is illuminating with regard to the error involved in the approximation. We also observe that there is a fine dependency to order one on* $\chi$ *which may be computed once the through flow is known! The above set of equations gives a closed system for the functions governing the through flow up to* $O(\varepsilon^2)$ *in a direct problem, when given the geometric definition of the blades in a row. This theory of through flow in an axial turbomachine is not valid near the locus of the leading/trailing edges of the row.*

*The* need for *transmission conditions* between *external force-free (outside of the row) through-flow* and *internal (in the row) through-flow with the fictive source force* **F** *appears readily as soon as any numerical treatment of the whole through flow is attempted in a two-row stage!* To *zeroth order,* these *transmission conditions,* found by matching these two disconnected through flows, are rather simple and obvious on physical grounds! Indeed, they express the fact that *mass flow is conserved,* as well as the *component of momentum parallel to the leading/trailing edge.*

More precisely, *near the leading/trailing edges of a blade,* one must make a *local analysis,* by setting [see (8.11f)].

$z = h(r) + \varepsilon z^*$, with $\mathbf{v}^*_{k,0}$, according to the inner expansion (8.11e), and with (8.11f) instead of the *zeroth order equation* $\partial \underline{\mathbf{G}}_{k,0}/\partial\chi = 0$ of the hierarchy (8.11g). To *zeroth order, we get a system of equations for a cascade flow:*

$$\partial \mathbf{G}^*_{k,0}/\partial\chi + 2\pi r \partial J^*_{k,0}/\partial z^* = 0, \tag{8.14f}$$

but the configuration is that of a *semi-infinite cascade flow*, where

$$J^*_{k,0} = \mathbf{Z}^*_{k,0} - (dh/dr)\,\mathbf{R}^*_{k,0}. \tag{8.14g}$$

*In fact,* to *zeroth order, we simply write the following two transmission conditions:*

$$[[\mathbf{v}^*_{k,0}.\boldsymbol{\tau}]] = 0 \quad \text{and} \quad [[2\pi r \cos\alpha\,\Delta^*\mathbf{v}^*_{k,0}.\boldsymbol{\mu}]] = 0, \tag{8.14h}$$

where $\tau$ is the unit vector tangent to the curve $\Gamma$ defined by $\{z = h(r), \theta = \Theta(r, h(r))\}$, and $\boldsymbol{\mu} = \boldsymbol{\tau} \wedge \boldsymbol{\nu}$, where $\boldsymbol{\nu}\,.\,\boldsymbol{\tau} = 0$. In (8.14h), $[[U]] = U_{+0} - U_{-0}$ is the *jump in U* at the leading/trailing edges of a row, and $\Delta^* = 1$, for $z^* < 0$ and $\Delta^* = \chi_{i,0} - \chi_{e,0}$, for $z^* > 0$.

Finally, for the angle $\alpha$, we have the following relation:

$$\tan\alpha = -r[\partial\Theta/\partial r + (dh/dr)\partial\Theta/\partial z]_{z=h(r)}. \tag{8.14i}$$

J.P. Veuillot, who defended his doctoral thesis in engineering in *1974*, chose my "*two stream function approach*" *(1966),* working on turbomachinery flow in the context of the G-Z theory at ONERA (France). He published two papers in 1971, where some numerical results are given for the "*inverse problem*" (assuming the "*circulation*" to be assigned) and for the "*direct problem*" (assuming a known "*blade row geometry*").

## 8.3 Asymptotic Modelling of Rolled-Up Vortex Sheets

By definition, a free vortex sheet is a regular discontinuous surface in the fluid domain, across which the tangential velocity component, but not the normal velocity component, is discontinuous. The vortex sheet behind a wing is represented schematically in Fig. 8.2 (*reproduced from p. 222 of* [13], cited in "Introduction").

### *8.3.1 Some Preliminary Comments*

*A rolled vortex sheets* is a region where the contiguous branches of the rolled sheet are so close to each other that they are very difficult to capture by numerical simulation (see pp. 203–230 in [13], cited in "Introduction"). The flow is assumed irrotational, but its vorticity is concentrated on the sheet. The vortex filament with vorticity continuously spreading over the tubular region is a *physical model* of the rolled vortex sheet.

Let $h$ be the distance between two consecutive turns. Then if $\mathbf{n}$ is the unit vector normal to the sheet, a physical argument suggests that

$$\boldsymbol{\Omega} = (\mathbf{n} \wedge [\mathbf{u}])/h, \tag{8.15a}$$

where $[\mathbf{u}]$ is the jump in the velocity $\mathbf{u}$ across the rolled vortex sheet, which should be a *good approximation* for the *vorticity spread over the vortex filament which models the rolled sheet*.

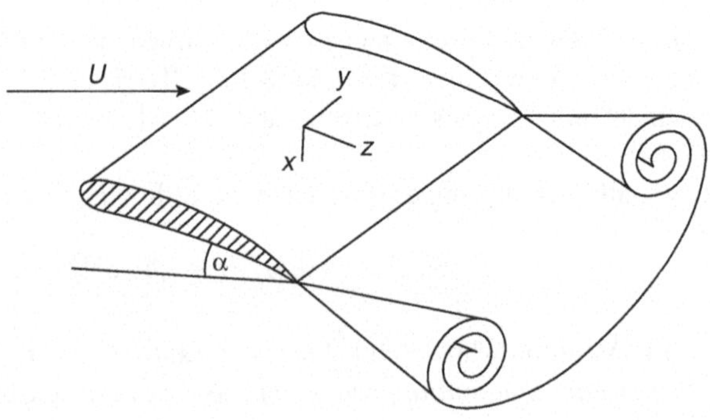


**Fig. 8.2** Schematic representation of a vortex sheet behind a wing

Such an argument was formalized in *1977* with *Guiraud in ONERA (France).* In quite a convincing, but not strictly rigorous way, it relies on an algorithm that allows one to relate a *continuous rotational flow whose vorticity is spread over a region to a corresponding irrotational discontinuous flow whose vorticity is concentrated on a rolled sheet.* Of course, this correspondence holds only in the *asymptotic limit* when *the turns of the sheet are infinitely close to each other* (see Fig. 8.3a).

In *G-Z theory, two parameters* play a major role. *One* is the *slenderness parameter,* which for the *leading-edge conical sheet is the distance to the focus of the spiral* as, in Fig. 8.3a. The *second* is the *closeness parameter,* the reciprocal of the number of turns or the distance between turns. For the conical leading-edge vortex sheet, it turns out that the *second small parameter* is of the order of the *square of the first,* if we adopt the useful convention that the logarithms are of order one. It is through the slenderness parameter that the core expansion is influenced by the exterior solution, so this takes into account the departure from circular symmetry.

*For the leading-edge core, the very fact that closeness is slenderness squared means that the departure from circular symmetry should occur at the order after the one computed by Mangler and Weber*, and we observe that, through a careful

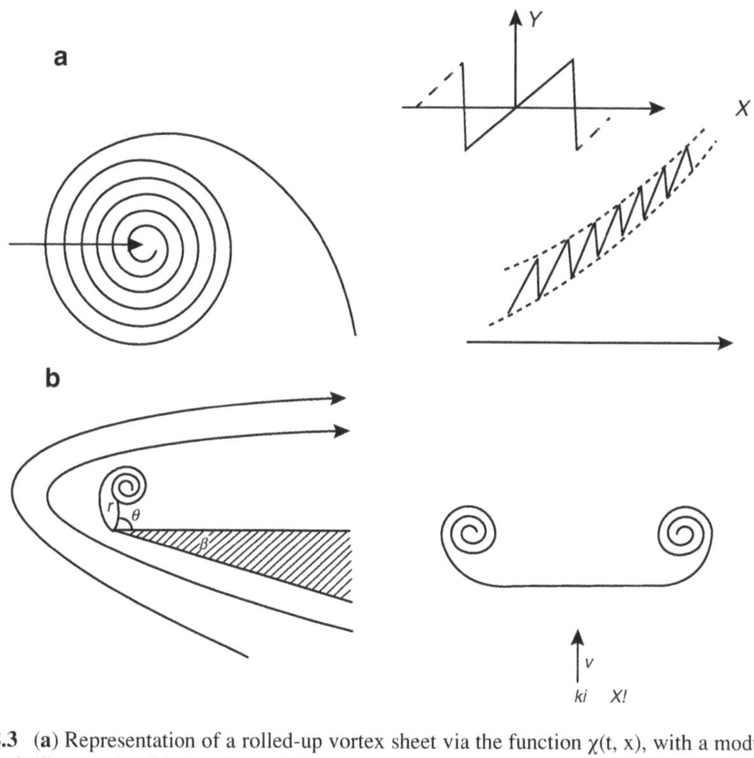


**Fig. 8.3** (**a**) Representation of a rolled-up vortex sheet via the function $\chi(t, x)$, with a modulated sawtooth-like graph; (**b**) Accelerated flow past an edge; similar flow for an extended plate. Reproduced from [13], cited in "Introduction": (**a**) p. 225, and (**b**) p. 228

inspection of the formulae derived by *Mangler & Weber, strongly deconstructed by Guiraud in <2>, it is apparent that their solution has a double scale structure!* It was to *satisfy our curiosity* that we computed the part of this corresponding to closeness, including it in our 1977 paper (with Guiraud), where two applications—to the *core of a leading edge* vortex and also to the famous *Kaden problem*—are performed (see J. Fluid Mechanics, vol. 79, part 1, pp. 93–112).

Indeed, it was thanks to the insistence of one referee, who was not satisfied (?) by our argument that *ellipticity of the spiral is related to the slenderness expansion,* and that it would be worthwhile having a separate process in order to deal with the closeness expansion, that we worked out Kaden's problem (see pp. 104–110 in our *above-mentioned 1977 paper with Guiraud*). In *1931,* Kaden obtained an asymptotic representation of the core of the semi-infinite sheet emerging from the potential flow around a flat edge which is instantaneously removed (see Fig. 8.3a), and this provides a model for the rolling-up of a trailing vortex sheet as described in *Moore and Saffman.*

In the framework of the *G-Z theory, various papers were published in the years 1977–1982. For example,* the *compressible inviscid case* was considered in a paper in *1980,* and in a *review* in 1982 <3>. A *mechanism of instability* was considered in *1979*, and in a second paper, in *1979,* it was shown that, concerning the *viscous diffusion of rolled* and *tightly wound vortex sheets*, the *viscous diffusion* may be described by a *linear heat equation.*

We found, with Guiraud, that the following very practical observation could motivate our asymptotic approach: a*symptotic modelling can be substituted for numerical simulation precisely when this simulation becomes so stiff, due to close spacing, that it is increasingly difficult to continue it when the vorticity is concentrated on a sheet.* A first application of this G-Z approach may be found in the paper by *Huberson. This is a good example of interdisciplinarity between the "numerical" approach and our RAM approach !*

## 8.3.2 Double Scale Structure

The rolled-up vortex sheet is represented by the function $\chi(t, x)$ with a modulated *sawtooth-like graph* (see Fig. 8.3a) so that the overall equation of the many sheets is

$$\chi(t,\mathbf{x}) = (2k+1)\pi,\ \ k = \ldots -2,\ -1, 0, 1, 3, \ldots, \tag{8.15b}$$

and the closeness assumption is expressed by

$$|\partial\chi/\partial t|, |\nabla\chi| >> 1, \tag{8.15c}$$

in such a way that the whole sheet is in fact given by $\chi =$ constant. This is multivalued, and in local cylindrical coordinates (r, $\phi$, $\sigma$), we write $\chi = \phi + F(t, r, \sigma)$, whence the right-hand side of (8.15b) expresses the multiple valuedness of $\phi$.

If we assume, according to the technique of multiple scaling and averaging, that the given time-dependent incompressible irrotational flow has many closely spaced vortex sheets, each carrying a weak vorticity, using (8.15b) and (8.15c), and set for $\mathbf{u}$ and $p$ the functions $\mathbf{u}^*(\mathrm{t}, \mathbf{x}; \chi(\mathrm{t}, \mathbf{x}))$ and $p^*(\mathrm{t}, \mathbf{x}; \chi(\mathrm{t}, \mathbf{x}))$, then we get as a starting system of equations:

$$\begin{aligned}
&\nabla\chi.\partial\mathbf{u}^*/\partial\chi + \nabla.\mathbf{u}^* = 0,\\
&(\partial\chi/\partial\mathrm{t} + \mathbf{u}^*.\nabla\chi)\partial\mathbf{u}^*/\partial\chi + (\partial p^*/\partial\chi)\nabla\chi + \partial\mathbf{u}^*/\partial t\\
&+(\mathbf{u}^*.\nabla)\mathbf{u}^* + \nabla p^* = 0,\\
&\nabla\chi \wedge \partial\mathbf{u}^*/\partial\chi + \nabla\wedge\mathbf{u}^* = 0,
\end{aligned} \tag{8.15d}$$

$$(\partial\chi/\partial\mathrm{t} + \mathbf{u}^*.\nabla\chi) = 0,\ \ \chi = (2\mathrm{k}+1)\pi, \tag{8.15e}$$

$$[p^*] = \nabla\chi\,.\,[\mathbf{u}^*] = 0 \text{ across } \chi = (2\mathrm{k}+1)\pi. \tag{8.15f}$$

In fact, $\mathbf{u}^*$ has a *sawtooth appearance*, as a function of $\chi$, and the way to *enforce the absence of secular terms* is to demand that $\mathbf{u}^*$ and $p^*$ be a function of $\chi$. We then observe that, on $\chi = (2\mathrm{k}+1)\,\pi$, $\mathbf{u}^*$ *but not $p^*$ suffers a discontinuity!*

### 8.3.3 *Taking Advantage of (8.15b,c) in Order to Solve the System (8.15d) with (8.15e) and (8.15f)*

We solve approximately, through an expansion process, assuming that, when applied to the flow variables, $\partial\chi/\partial\mathrm{t}$ and $\nabla$ are of order $(1/s)$, where $s$ is the *slenderness parameter*. In the application to the leading-edge core, $s$ is found to be, $O(\theta\,|\log\theta|)$, where $\theta$ is the angular distance to the focus of the spiral (on a unit sphere). We assume further that $|\partial\chi/\partial\mathrm{t}|$ and $|\nabla\chi|$ are of order $(1/c)\,(1/s)$, where $c$ is the closeness parameter, which is supposed small.

First, let us consider the third equation in system (8.15d), viz.,

$$\nabla\chi\wedge\partial\mathbf{u}^*/\partial\chi + cs\nabla\wedge\mathbf{u}^* = 0, \tag{8.15g}$$

and then expand $\mathbf{u}^*$, $p^*$, and $\chi$ according to

$$\begin{aligned}
U^* &\equiv (\mathbf{u}^*, p^*, \chi) = U^{*(0)} + \mathrm{c}\,U^{*(1)} + \mathrm{c}^2 U^{*(2)} + \dots\\
&= U_0^* + \mathrm{c}\,U_1^* + \mathrm{c}^2\,U_2^* + \dots\,.
\end{aligned} \tag{8.15h}$$

Instead of introducing the small parameter $c$ into the equations, we leave them as they stand and solve them iteratively in the obvious manner. As a matter of fact, in the application, the expansion with respect to c is a coordinate expansion and its structure is obvious from the results.

To the *leading approximation*, the model flow satisfies the continuity and Euler equations

$$\nabla.\mathbf{u}_O{}^* = 0, \partial \mathbf{u}_O{}^*/\partial t + (\mathbf{u}_O{}^*.\nabla)\mathbf{u}_O{}^* + \nabla p_O{}^* = 0, \tag{8.15i}$$

with

$$(\partial \chi_O/\partial t + \mathbf{u}_O{}^*.\nabla \chi_O) = 0, \tag{8.15j}$$

which is deduced from (8.15e).

Proceeding to the *second approximation*, we find

$$\nabla \chi_O.\partial \mathbf{u}_1{}^*/\partial \chi = 0, \nabla \chi_O \wedge \partial \mathbf{u}_1{}^*/\partial \chi = -\nabla \wedge \mathbf{u}_O{}^* = 0, \tag{8.15k,1}$$

from which

$$\begin{aligned} &\mathbf{u}_1{}^* = \underline{\mathbf{u}}_1{}^* + \{[\nabla \chi_O \wedge \omega_O{}^*]/|\nabla \chi_O|^2\}\mathrm{Y}_1(\chi), \\ &\omega_O{}^* = \nabla \wedge \mathbf{u}_O{}^*, \\ &p_1{}^* = \underline{p}_1{}^*, \end{aligned} \tag{8.15m}$$

where

$$\mathrm{Y}_1(\chi) + \chi \text{ for } |\chi| < \pi, \mathrm{Y}_1(\chi) - \text{periodic with period } 2\pi, \tag{8.15n}$$

and we observe that the above $\underline{\mathbf{u}}^*$ and $\underline{p}^*$ are obtained in the process of going from any $\mathbf{u}^*$ and $p^*$ *to* $\underline{\mathbf{u}}^*$ and $\underline{p}^*$ by the *averaging process*.

*Then,* from *periodicity of* $\mathbf{u}^*$ and $p^*$, we get the set of *averaged equations*, which bear (only!) some analogy with the usual equations of incompressible flow. We find easily

$$\partial \mathbf{u}_0{}^*/\partial \chi = \partial p_0{}^*/\partial \chi = 0, \text{ giving } \underline{\mathbf{u}}_0{}^* = \mathbf{u}_0{}^*, \ \underline{p}_0{}^* = p_0{}^*. \tag{8.15o}$$

It is interesting to note that (8.15k) and (8.15l) are *soluble* with respect to $\partial \mathbf{u}_1{}^*/\partial \chi$ *if and only if* $\nabla \chi_O \cdot \boldsymbol{\omega}_O{}^* = 0$. From the above-mentioned *averaged equations where* $\underline{\mathbf{u}}^*$ and $\underline{p}^*$ *are present,* and *bearing some* analogy with the usual equations of incompressible flow, we obtain:

$$\begin{aligned} &\frac{\partial \mathbf{u}_1{}^*}{\partial t} + (\mathbf{u}_O{}^*.\nabla)\underline{\mathbf{u}_1{}^*} + \left(\underline{\mathbf{u}_1{}^*}.\nabla\right)\mathbf{u}_O{}^* + \nabla \underline{p_1{}^*} = 0, \\ &\nabla.\overline{\mathbf{u}_1{}^*} = 0, \\ &\partial \chi_1/\partial t + \mathbf{u}_O{}^*.\nabla \chi_1 + \underline{\mathbf{u}_1}{}^*.\nabla \chi_O = 0. \end{aligned} \tag{8.15p}$$

### *8.3.4 Dynamical Condition on the Sheet*

For this one must examine the *third approximation*! From (8.15d), taking the two equations (8.15j) and (8.15k) into account, where $\mathbf{u}_2{}^*$ and $p_2{}^*$ *are present, and provided that*

$$\nabla\chi_1.\boldsymbol{\omega}_O^* + \nabla\chi_0.\nabla \wedge \underline{\mathbf{u}_1^*} = 0, \tag{8.15q}$$

we get the expressions for $\mathbf{u}_2^*$ and $p_2^*$. However, the *third of the three above-mentioned equations* is *soluble for* $\partial p_2^*/\partial\chi$ if and only if

$$\partial\Gamma_O/\partial t + \mathbf{u}_O^*.\nabla_T\Gamma_O = 0, \tag{8.15r}$$

where $\Gamma$ is the *jump in the velocity potential across the sheet and* $\nabla_T$ the gradient operator along it. We then observe that, as a consequence of irrotationality for the exact solution, one has $[\mathbf{u}] = \nabla_T\Gamma$. Of course, (8.15r) is a *well-known condition in the theory of vortex sheets* and it is not at all surprising that we should recover it!

For the conical vortex sheet and application to the *core of a leading-edge vortex considered in § 4 of our 1977 paper with Guiraud in JFM* (where a *second application* to *the Kaden problem* is also considered), *it is found that* (8.15r) is *well satisfied automatically*!

The most fruitful continuation of our G-Z work would be to use it to devise an efficient numerical process for computing complex flow configurations involving rolled-up vortex sheets as in *Rebach* (from ONERA) in *1975*.

### *8.3.5 Vortex Sheets and Related Flow Structure*

In "*Vortex Motion*", eds. H.G Hornung/E.-A Müller, 1982 <3>, the reader will find an asymptotic model, which concerns a situation in which *there are many closely spaced vortex sheets*. We may mention two flow configurations in which this occurs. The *first* is the *core of a highly rolled-up vortex sheet*, while the *second* concerns the set of trailing vortex sheets which are formed at the *trailing edges of the blades arranged in a row in an axial turbomachine* (see Sect. 8.2).

We assume that there is a *small parameter* built into the flow, viz., t*he ratio of the spacing between two consecutive sheets to the width of the region covered by the sheets*! We set *C* for this parameter and, as in our *1979 paper* (in J. Fluid Mechanics, 90, pp. 197–201), we call it the "*closeness parameter*". The purpose of our asymptotic modelling with Guiraud in this situation was to derive a model which avoided the stiff problem of computing the flow with a numerical code capable of capturing many sheets!

In fact, the problem is one of *multiple scaling*, in the terminology of asymptotics. A version, especially suited to this kind of problem was devised by Guiraud and Zeytounian in the previous papers (1977 and 1979), with the purpose of describing rolled vortex sheets. But its scope is more general, as can be seen from the analysis given in the above-mentioned 1982 paper [26] (see pp. 35–37), cited in "Introduction".

A basic result when *modeling rolled vortex sheets* is the following system of equations for the two functions $\mathbf{V} = \mathbf{k}_0 \wedge \mathbf{u}_0^*$ and $\Sigma = S_1^*$:

$$\partial \mathbf{V}/\partial t + \mathbf{u}_0{}^*.\nabla\mathbf{V} - \mathbf{V}.\nabla\mathbf{u}_0{}^* + (\nabla.\,\mathbf{u}_0{}^*)\mathbf{V}$$
$$+ \left\{[\mathbf{k}_0.(\nabla \wedge \mathbf{u}_0{}^*)]/|\mathbf{k}_0|^2\right\}(\mathbf{k}_0 \wedge \mathbf{V})$$
$$+ (1/\gamma M^2)\{[\mathbf{k}_0 \wedge (\nabla p_0{}^*)]/\gamma \rho_0{}^*\}\Sigma = 0, \tag{8.16a}$$

$$\partial\Sigma/\partial t + \mathbf{u}_0{}^*.\nabla\Sigma + \left\{[\mathbf{k}_0 \wedge (\nabla_0{}^*)]/|\mathbf{k}_0|^2\right\}.\mathbf{V},$$

with

$$\partial \rho_0{}^*/\partial\chi = \partial S_0{}^*/\partial\chi = 0. \tag{8.16b}$$

The first of these equations, viz., (8.16a) for **V**, is a consequence of the equation of *vorticity. The* important point is that $\chi$ *does not occur in* (8.16a). *In fact*, these equations constitute a set of ordinary differential equations along the trajectories of the velocity fields at the zeroth order of approximation $\mathbf{u}_0{}^*$. One may set

$$(\mathbf{V}, \Sigma) = F_0(t^\circ,\, t)\, (\mathbf{V}^\circ, \Sigma^\circ), \tag{8.16c}$$

stating that the fields $\{\mathbf{V}(t, \mathbf{x}), \Sigma(t, \mathbf{x})\}$ are known at time t once the fields $\{\mathbf{V}^\circ(\mathbf{x}), \Sigma^\circ(\mathbf{x})\}$ are *given* at time t°. In principle, the linear operator $F_0(t^\circ, t)$ can be computed from a numerical code when the zeroth order field $\mathbf{u}_0{}^*$, $p_0{}^*$, $\rho_0{}^*$ is *known.* We may set

$$(\mathbf{k}_0 \wedge \mathbf{u}_1{}^*, S_1{}^*) = F_0(t^\circ,\, t)\, (\mathbf{k}_0 \wedge \mathbf{u}_1{}^{*\circ}, S_1{}^{*\circ}), \tag{8.16d}$$

and the whole of the dependency on $\chi$ appears through $\mathbf{u}_1{}^{*\circ}$and $S_1{}^{*\circ}$, the values of $\mathbf{u}_1{}^*$ and $S_1{}^*$ at t°. A number of configurations are allowed with this description in the our above-mentioned 1982 paper [26] (see pp. 37–46), cited in "Introduction". But here we end our discussion of vortex flows!

*We hope that our point of view with Guiraud, illustrated in our 1982 paper <3>, will convince some readers that asymptotic RAM techniques provide very powerful tools for the purpose of modelling problems which are too stiff for pure numerical simulation/computation. I am sure that, the more efficient computing becomes, the more we shall need techniques capable of unravelling stiff fluid dynamics problems, and the RAM approach, among others, will prove to be a very effective support/tool for this purpose!*

## 8.4 Long Nonlinear Surface Waves on Water and Soliton

### 8.4.1 Dynamics: An Overview

One of the most interesting and successful applications of *nonlinear hydrodynamics* is to the *wave motion, under gravity, of water with a free surface*. Studies on water waves have always been enriched by cross-fertilization among diverse fields of science, including applied mathematics and singular perturbation/asymptotic techniques.

In "*Theory and Applications of Non-viscous Fluid Flow*", *2002* [19, pp. 73–81], cited in "Introduction", and also in "*Asymptotic Modelling of Fluid Flow Phenomena*", FMIA 64, 2002 [13, pp. 147–165], cited in "Introduction", the reader can find a *short account* of the adventure of long n*onlinear surface waves* from *Boussinesq, Cauchy, Poisson, and Stokes* to *Craik, Davey-Stewartson, Kadomtsev-Petviashvili, Newell, Zabusky-Kruskal, Zakharov*. *Then Zeytounian's review paper of 1995 was dedicated by the Editorial Board of Uspekhi Fizicheskikh Nauk* (*UFN*—Journal of the Russian Academy of Sciences) to the *30th anniversary of the publication of the paper by Zabusky and Kruskal, in which the term "soliton" was first mentioned in the scientific literature*. Indeed, the theory of solitons is *attractive* and *exciting*—it brings together many branches of mathematics, some of which touch upon *profound ideas*. Several of its aspects are amazing and beautiful, and the book by *Newell* [5], cited in Chap. 1, is interesting and pertinent here, and strongly recommended!

### 8.4.2 *Pivotal Dimensionless Problem and the 2D Single Equation à la Boussinesq for Long Waves in Shallow Water*

In the dimensionless *pivotal starting problem* (8.17a), in the condition *(d)*, T* is the dimensionless Bond-Weber parameter which takes into account the dynamic influence of the *surface tension on the free upper surface* $z=\varepsilon\,\zeta(t, x, y)$. *Condition (b)* is only written for a *flat bottom*, but the reader can find the case of an uneven bottom in our 1995 paper, published in *Russian Physics-uspekhi* (English ed. V.**38** (12), 1333–1381). It is often assumed that $T^*=0$.

For the dimensionless *velocity potential* $\phi(t, x, y, z)$, we have the following so-called *pivotal starting mathematical problem* (see <4>):

$$
\begin{aligned}
&(a)\ \phi_{zz}+\delta^2\phi_{xx}+\Delta^2\phi_{yy}=0 && \text{for } -1\le z<\varepsilon\,\zeta(t,x,y);\\
&(b)\ \phi_z=0 && \text{on } z=-1;\\
&(c)\ \phi_z=\delta^2\zeta_t+\varepsilon\left[\delta^2\phi_x\,\zeta_x+\Delta^2\phi_y\,\zeta_y\right] && \text{on } z=\varepsilon\,\zeta(t,x,y);\\
&\qquad \phi_t+(1/2)\left[\varepsilon\,(\phi_x)^2+\varepsilon\,(\Delta/\delta)^2\left(\phi_y\right)^2+\left(\varepsilon/\delta^2\right)(\phi_z)^2\right]+\zeta(t,x,y)\\
&(d)\ =\delta^2T^*\left[\zeta_{xx}+(\Delta/\delta)^2\zeta_{yy}-(3/2)\,(\varepsilon\,\delta)^2(\zeta_x)^2\zeta_{xx}\right.\\
&\qquad\qquad \left.+\,O\left(\varepsilon^2\Delta^2\right)\right] \text{ on } z=\varepsilon\,\zeta(t,x,y).
\end{aligned}
\tag{8.17a}
$$

In (8.17a), we assume that $\varepsilon\ll 1$, $\delta\ll 1$, and $\Delta\ll 1$, and in the framework of our RAM approach, two similarity relations:

$$\delta^2 = \kappa^\circ\varepsilon \text{ and } \Delta = \nu^\circ\,\varepsilon, \text{where } \kappa^\circ \text{ and } \nu^\circ \text{ are } 0(1) \text{ when } \varepsilon \to 0. \tag{8.17b}$$

The Laplace equation (a) of (8.17a) is the *only equation which contains z in its solution,* and the variation in z may be made explicit by formally expanding its solution in powers of $\delta^2$ and $\Delta^2$, viz.,

$$\phi = \phi_{00} + \delta^2\phi_{10} + \delta^4\phi_{20} + \Delta^2\phi_{01} + \delta^6\phi_{30} + \delta^2\Delta^2\phi_{11} + \ldots, \tag{8.17c}$$

assuming that

$$\phi_{00} = F(t, x, y) \tag{8.17d}$$

*is the unknown value of the velocity potential* $\phi$ on $z = -1$!

A rather simple calculation leads, via the Taylor expansions and taking into account the boundary conditions on $z = \varepsilon\,\zeta(t, x, y)$, to a single equation for the function

$$F^* = F0 + \varepsilon F1\,, \tag{8.17e}$$

*when we observe that the function F(t, x, y) in* (8.17d) *is obviously an implicit function* of $\varepsilon$, $F = F0 + \varepsilon F1 + \varepsilon^2 F2 + \ldots$.

The single equation (8.17f) for $F^*$ is a *generalized quasi-one-dimensional Boussinesq* equation, discovered in 1991 <5>, for *nearly two-dimensional long waves in a shallow water:*

$$\partial^2F^*/\partial t^2 - \partial^2F^*/\partial x^2 - \varepsilon\left[(\nu^\circ)^2/\kappa^\circ\right]\partial^2F^*/\partial y^2 + \varepsilon\{[\partial F^*/\partial x]^2 + (½)[\partial F^*/\partial t]^2 - (1/3)\varepsilon\,\kappa^\circ\left[\partial^4F^*/\partial x^2\partial t^2\right] = 0. \tag{8.17f}$$

## 8.4.3 Solitary and Cnoidal Waves

In the case of a *single one-dimensional Boussinesq* equation, when $(\nu^\circ)^2 = 0$ in (8.17f), if $F^* = \Phi(\xi)$ with $\xi = x - ct$, we obtain for the corresponding *free surface position* (profile):

$$\zeta = \text{sech}^2\left\{(3/2\kappa^\circ)^{1/2}(\xi - \xi_0)\right\}, \tag{8.18a}$$

which is a *solitary hill* with a single crest at $\xi = \xi_0$, whose amplitude diminishes to zero as $|\xi| \to \infty$ was discovered by *Scott Russell in 1834*. But besides the above solitary wave, *periodic permanent waves (referred to as cnoidal)* are also possible in the framework of the single one-dimensional Boussinesq equation.

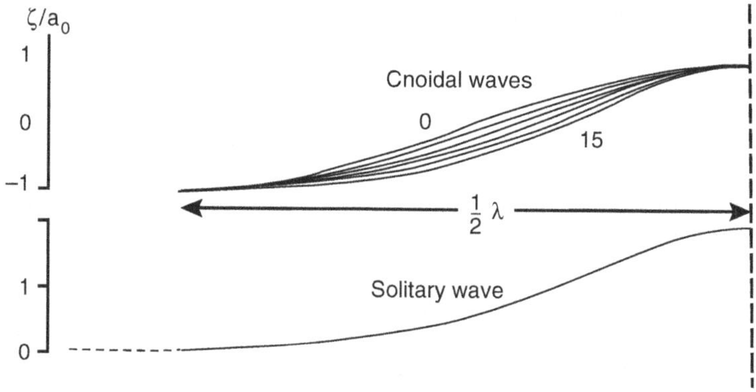


**Fig. 8.4** Waveforms for cnoidal waves

We observe that the above wave is periodic, so we can now have a train of periodic waves on shallow water. In particular, the wave train following an undular bore can be regarded as a train of *cnoidal waves.*

The cnoidal waveforms are plotted in Fig. 8.4 (reproduced from *p. 99 of* our *unpublished typescript <4> 1993).* The *solitary wave* is the *limiting form of the cnoidal wave with infinite wavelength.*

### 8.4.4 KdV and KP Equations for Weakly Nonlinear Long Waves in Shallow Water

In *1895,* Korteweg and de Vries published a paper entitled *"On change of form of long waves advancing in a rectangular canal, and a new type of long stationary waves". Below* we show that the *KdV equation emerges* very naturally and in a *consistent way via asymptotics with respect to* $\varepsilon \ll 1$, *when we start from the following free surface dimensionless problem:*

$$\begin{aligned}
&(a)\ \ \phi_{zz} + \varepsilon\phi_{\xi\xi} = 0, \ \text{when}\ 0 < z < 1 + \zeta(\tau,\xi),\\
&\quad\ \ \phi_z = 0, \ \text{on}\ z = 0,\\
&(b)\ \ \phi_z = \varepsilon^{3/2}\zeta_\tau + \varepsilon\,\phi_\xi\zeta_\xi - \varepsilon^{1/2}\zeta_\xi,\\
&(c)\ \ \varepsilon^{3/2}\phi_\tau + \zeta + (1/2)\left[\varepsilon\left(\phi_\xi\right)^2 + (\phi_z)^2 - \varepsilon^{1/2}\phi_\xi\right] = 0,\\
&(d)\ \ \text{on}\ z = 1 + \varepsilon\,\zeta_1(\tau,\xi) + \varepsilon^2\zeta_2(\tau,\xi)\,,
\end{aligned} \tag{8.18b}$$

with the new variables $\xi = \varepsilon^{1/2}(x - t)$ and $\tau = \varepsilon^{3/2}\,t$, where $\varepsilon = (h_0/\lambda_0)^2 \ll 1$ and $\lambda_0$ is the wavelength of the long waves.

Taking into account the following asymptotic representations for $\phi$ and $\zeta$:

$$\begin{aligned}&(e)\quad \phi = \varepsilon^{1/2}\{\phi_1(\tau,\ \xi, z) + \varepsilon\, \phi_2(\tau,\ \xi, z)\ +\ \varepsilon^2\phi_3(\tau,\ \xi, z)\ + \ldots\},\\ &(f)\quad \zeta = \varepsilon\, \zeta_1(\tau, \xi)\ + \varepsilon^2\zeta_2(\tau, \xi) + \ldots,\end{aligned} \tag{8.18c}$$

at the free surface (c) in (8.18b), the derivatives $\phi_\tau$, $\phi_\xi$, and $\phi_z$ may be expressed by means of the Taylor expansions, and by substitution we obtain in place of (8.18c), to different orders in $\varepsilon$, respectively:

$$\begin{aligned}&\phi_1 = F(\tau, \xi) \text{ and } \zeta_1 = F_\xi,\ \ \phi_2 = -\ (½)\ z^2F_{\xi\xi} + G(\tau, \xi),\\ &\zeta_2 - (\phi_2)_\xi + (½)\ (\zeta_1)^2 + (\phi_1)_\tau = 0, \text{ on } z = 1,\\ &\phi_3 = (1/4!)\ z^4F_{\xi\xi\xi\xi} - (½!)\ z^2G_{\xi\xi} + H(\tau, \xi).\end{aligned} \tag{8.18d}$$

But at order $\varepsilon^2$, from the free surface condition $\phi_z = 0$, on $z = 0$, we also derive the following relation between $\phi_1$, $\phi_2$, $\phi_3$, and also $\zeta_1$ and $\zeta_2$ :

$$\left[(\phi_3)_z\right]_{z=1} + \zeta_1\left[(\phi_2)_{zz}\right]_{z=1} - (\zeta_1)_\xi\left[(\phi_1)_\xi\right]_{z=1} = (\zeta_1)_\tau - (\zeta_2)_\xi. \tag{8.19}$$

As direct consequence of this last relation (8.19), when we use the above solutions for $\phi_1$, $\phi_2$, $\phi_3$, and also the above relation $(\zeta_2)_{z=1}$ in (8.18d), we derive the following *reduced KdV equation* for $\zeta_1$:

$$\partial\zeta_1/\partial\tau + (3/2)\zeta_1\partial\zeta_1/\partial\xi + (1/6)\partial^3\zeta_1/\partial\xi^3 = 0. \tag{8.20a}$$

This is in fact a *compatibility condition* for the consistency of our asymptotic derivation connected with (f) and (e) in (8.18c). It admits only wave solution *moving to the right*.

Interest waned after the solution of the famous *Airy-Stokes* paradox (see pp. 91–95 in our companion *2014* book) by *Boussinesq* and *Rayleigh* and was sporadic prior to *Zabusky* and *Kruskal*'s discovery that the solitary waves—which they called "*solitons*"—typically dominate the asymptotic solution of the KdV equations.

As a simple canonical KdV equation, one can consider

$$\partial u/\partial t + u\partial u/\partial x + \kappa^2\, \partial^3 u/\partial x^3 = 0, \tag{8.20b}$$

with

$$u(0,\ x) = \cos \pi x,\ 0 < x < 2, \tag{8.20c}$$

and conditions

$$u,\ \partial u/\partial x\,,\ \partial^2 u/\partial x^2 \text{ all periodic on } [0,\ 2] \text{ for all t.} \tag{8.20d}$$

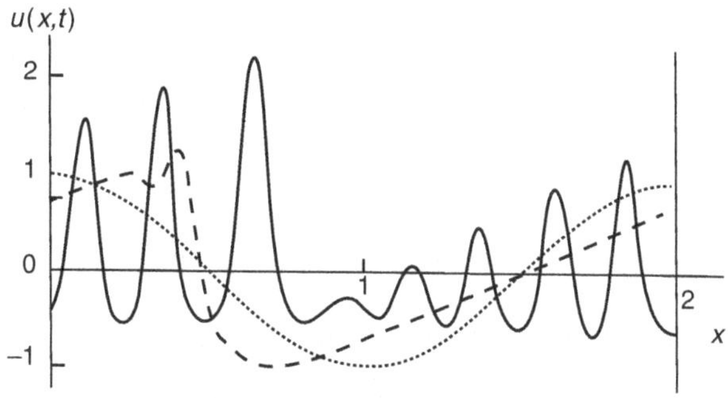


**Fig. 8.5** Solution of the periodic boundary-value problem for the KdV equation (8.20b), with (8.20c) and (8.20d)

A set of results for the boundary-value problem (8.20b) with (8.20c) and (8.20d) is given in Fig. 8.5 (*reproduced from p. 79 of our book* [13], cited in "Introduction"), showing the initial profile for u at $t=0$ (*dotted line*), the profile at $t=1/\pi$ (*broken line*), and the profile at $t=3.6/\pi$ (*full line*).

Naturally, when the nonlinear surface waves in *weakly dispersing shallow water are not strictly one-dimensional*, the *KdV equation no longer applies!* One must derive a *new approximate model equation* for this case, the so-called *KP equation*! In fact, the asymptotic derivation of this *Kadomtsev-Petviashvili equation (1970)* closely follows that of the KdV equation, so we merely review the main points of this asymptotic derivation here.

First, as above, we use the new variables $\xi=\varepsilon^{1/2}(x-t)$ and $\tau=\varepsilon^{3/2}t$, z, and also $\eta=\varepsilon y$, in place of the above KdV variables. We thereby obtain the following dimensionless problem:

$$
\begin{aligned}
&(a)\quad \phi_{zz}+\varepsilon\phi_{\xi\xi}+\varepsilon^2\phi_{\eta\eta}=0,\\
&(b)\quad \phi_z=0,\ \text{on } z=0,\\
&(c)\quad \phi_z=\varepsilon^2\phi_\eta\,\zeta_\eta+\varepsilon^{3/2}\zeta_\tau+\varepsilon\,\phi_\xi\,\zeta_\xi-\varepsilon^{1/2}\zeta_\xi,\\
&\qquad\ (1/2)\varepsilon^2\left(\phi_\eta\right)^2+\varepsilon^{3/2}\phi_\tau+\zeta+(½)\left[\varepsilon\left(\phi_\xi\right)^2\right.\\
&(d)\quad \left.+\left(\phi_z\right)^2-\varepsilon^{1/2}\phi_\xi\right]=0,\ \text{with}\\
&(e)\quad \text{on } z=1+\varepsilon\,\zeta_1(\tau,\xi,\eta)+\varepsilon^2\zeta_2(\tau,\xi,\eta)+\ldots,
\end{aligned}
\qquad (8.21)
$$

where

$$\phi=\varepsilon^{1/2}\{F(\tau,\xi,\eta,z)+\varepsilon\,\phi_2(\tau,\xi,\eta,z)+\varepsilon^2\phi_3(\tau,\xi,\eta,z)+\ldots\}. \qquad (8.22)$$

From (8.21) (a)–(e) we see that the dependence on $\eta$ appears explicitly only at order $\varepsilon^2$, and as consequence the above results concerning the KdV equation remain unchanged up to $\phi_2$ in (8.18d). Instead of $\phi_3$ in (8.18d), we obtain the following solution:

$$\phi_3=(1/4!)\,z^4F_{\xi\xi\xi\xi}-(½!)\,z^2G_{\xi\xi}-(½)z^2F_{\eta\eta}+H(\tau,\xi,\eta), \qquad (8.23a)$$

$$\mathrm{F} = \mathrm{F}(\tau,\ \xi,\ \eta)\,\mathrm{and}\,\zeta_1 = \mathrm{F}_\xi. \tag{8.23b}$$

Now if we use (8.23a) for $\phi_3$, rather than the expression for $\phi_3$ in (8.19), we derive from the relation (8.19) the following new "two-dimensional" relation for $F(\tau, \xi, \eta)$:

$$(1/3)F_{\xi\xi\xi\xi} + F_{\eta\eta} + 3F_\xi F_{\xi\xi} + 2F_{\tau\xi} = 0. \tag{8.23c}$$

Finally, if we introduce the function:

$$f(\xi,\ \eta,\ \theta) = (2/3)F_\xi(\xi,\ \eta,\ 6\theta)\ \mathrm{with}\,\theta = \tau/6, \tag{8.23d}$$

we obtain the *classical KP equation* in the form

$$\partial/\partial\xi\left\{\partial f/\partial\,\theta + 6f\partial f/\partial\xi + \partial^3 f/\partial\xi^3\right\} + 3\partial^2 f/\partial\eta^2 = 0, \tag{8.24}$$

with

$$\zeta = (3/2)\varepsilon f + 0(\varepsilon^2).$$

The *N* soliton wave solution of the *KP equation* (8.24) was derived by *Satsuma* and its structure elucidated by *Miles*, but see also the useful review paper by *Freeman*. This term "*soliton*" emphasizes the *particle-like character* of these waves, *a train of eight* well-defined waves, which retain their identities in a collision (see Fig. 8.6, which is reproduced from p. 80 of our book [13], cited in "Introduction").

### 8.4.5 *Nonlinear Schrödinger (NLS) Equations and Schrödinger-Poisson (NLS-P) Equations*

Here the aim is to deduce the model equations governing a *slowly varying train of long surface waves in shallow water*, which are essentially sinusoidal and propagate in one direction! These model equations are derived via asymptotics, by applying

**Fig. 8.6** Three-soliton wave solution of the KP equation

the *method of multiple scales* and the *one-dimensional non-linear Schrödinger (NLS) equation*

$$-\,\mathrm{i}\partial \mathrm{A}/\partial \mathrm{t} + \partial^2\mathrm{A}/\partial \mathrm{x}^2 + \chi\, A|\mathrm{A}|^2 = 0, \tag{8.25}$$

to the water-wave problem. It was first derived for finite depths (classical problem) by Hasimoto and Ono in 1972. A similar *NLS equation* was deduced earlier, but for *an infinite depth*, by *Zakharov* [see (8.28) below].

For 2D surface water waves, instead of the *NLS equation* (8.25), *Benney* and *Roskes* and *Davey* and *Sterwatson* derived a system of two equations—a so-called, *NLS-Poisson model system.*

The reader can find a very similar system to (3.41) on p. 94 of Chap. 3 of our 2014 book:

$$-\,\mathrm{i}\partial \mathrm{A}/\partial \mathrm{t} + \lambda\, \partial^2\mathrm{A}/\partial \mathrm{x}^2 + \mu\, \partial^2\mathrm{A}/\partial \mathrm{y}^2 = \chi \mathrm{A}\, |\mathrm{A}|^2 + \delta\, \mathrm{A}\, \partial \mathrm{B}/\partial \mathrm{x}, \tag{8.26a}$$

$$\mathrm{a}\partial^2\mathrm{B}/\partial x^2 + \partial^2\mathrm{B}/\partial \mathrm{y}^2 = \mathrm{b}\partial\left(|\mathrm{A}|^2\right)/\partial \mathrm{x}. \tag{8.26b}$$

Figure 8.7 *(reproduced* from p. 194 of our *unpublished typescript <4> 1993)* is a schematic illustration of an *oblique envelope soliton* with $\alpha = 30°$. In fact, if we introduce the transformation

$$Z = \xi \cos \alpha + \eta \sin \alpha, \tag{8.27}$$

the *two-dimensional NLS equation* à la *Zakharov for the case of deep water* ($\rightarrow \infty$), viz.,

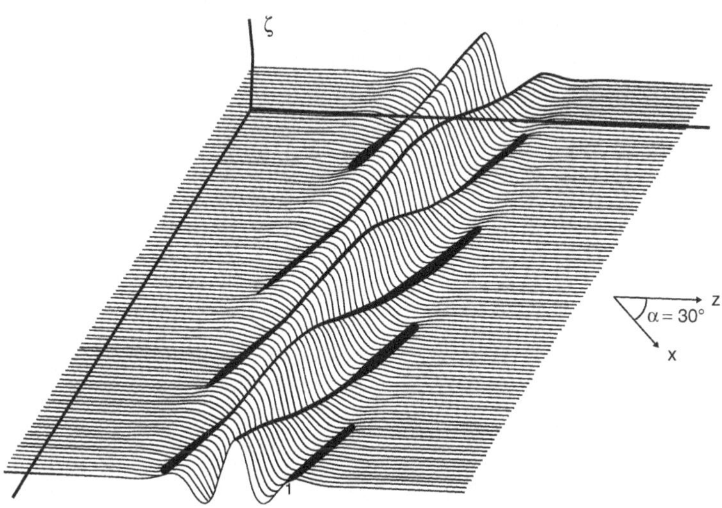


**Fig. 8.7** Sketch of a "three-dimensional" oblique plane envelope soliton

$$-\,i\partial A/\partial\tau + \partial^2 A/\partial\xi^2 - \partial^2 A/\partial\eta^2 + \beta A|A|^2 = 0, \qquad (8.28)$$

reduces to

$$-\,i\partial A/\partial\tau + \left[1 - 3\sin^2\alpha\right]\partial^2 A/\partial\zeta^2 + \beta A|A|^2 = 0, \qquad (8.29)$$

and any solution of equation (8.29) represents an oblique plane solution of (8.28), with $\alpha$ representing the angle between the propagation direction of the wave envelope and the propagation direction of the individual carrier wave.

For $\alpha < 35.26°$, equation (8.29) remains of the same type as the one-dimensional NLS equation (8.25), but for $\alpha > 35.26°$, the relative signs of the dispersive and the nonlinear terms change, and the properties of the solution also change substantially.

Expressions for the various constant coefficients λ, μ, χ, δ, *a*, and *b*, in (8.26a,b), for *capillary-gravity water waves* (when we take into account t*he surface tension* in the classical problem) are given by *Craik,* in Chap. 6 of his very interesting and well documented book *(1995).*

Figure 8.8 (reproduced from *p. 214 of our typescript, edited by University of Lille I, 1993 <4>*) shows the dependence of the coefficients in the NLS-P system of two coupled equations, with the influence of the surface tension via the *Bond–Weber number* T*, such that $T^{**} = \delta^2 T^*$. The various curves indicate where the coefficients change sign.

### 8.4.6 Some Final Remarks

We observe that, in the presence of a free surface, the vorticity of an inviscid incompressible fluid does not necessarily remain zero if it is zero initially! Indeed the free surface can intersect itself, as happens when a wave breaks and vortex

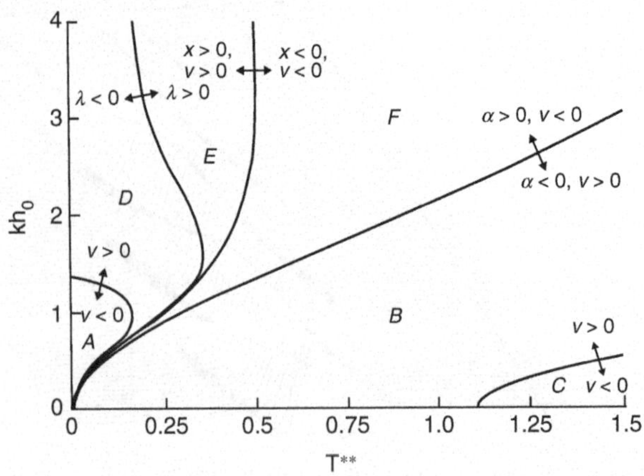


**Fig. 8.8** Dependence of coefficients in the NLS-P system on $kh_0$ and T**

sheets are formed! In such a case, instead of Laplace's equation, one must consider the full Euler equations for non-viscous incompressible water. It seems me that such a complete approach has rarely, if ever, been carried out!

In <4>, and in our *1995 review paper* <6> (*English translation of the Russian Uspekhi Fizicheskikh Nauk*), the reader can find a deep derivation of the various model equations mentioned above. It is important to understand that in some cases the establishment of these model equations from the above pivotal starting problem (8.17a)–(8.17e) is rather *"intuitive"* and "*heuristic*", and in fact it is not clear how to insert the relevant model equation into a hierarchy of rational approximations, which in turn emerge as a logical and valuable result from the exact mathematical formulation (8.17a)–(8.17e) chosen at the beginning!

## 8.5 Some Singular Aspects of the Blasius BL Problem Asymptotics of the Blasius BL Steady Incompressible 2D Problem

A basic problem in the theory of fluids with *small viscosity* $(Re \gg 1)$ concerns steady incompressible flow past a solid flat plate placed edgewise in a uniform stream. More precisely, the plate is understood to be a half-plane, say $y = 0$, such that $0 < x < \infty$, $-\infty < z < \infty$, and the 2D flow is assumed to approach a uniform flow far upstream, with velocity parallel to the plate and normal to its edge:

$$\lim_{x \to 0} \mathbf{u} = U\infty \mathbf{i} \text{ for all y , and z,} \tag{8.30}$$

where $U\infty = \text{const.} > 0$ and $\mathbf{i}$ denotes the unit vector in the direction of increasing x. Indeed, this is described in pp. 97–103 of our 2004 book [14], cited in "Introduction", and Fig. 8.9 (reproduced from p. 100 of the same book) shows the four regions needed for an asymptotic study of incompressible steady 2D fluid flow with vanishing viscosity.

The first region (I) is the "*Navier*" *leading-edge* region governed by the full steady 2D Navier equations, with the matching condition, when $x^* = Re\, x \to \infty$, which recovers the Blasius BL flow for the value $x = 0$, as required to compute the parabolic Prandtl-Blasius BL problem.

Region II is just the "intermediate" matching region governed by the above-mentioned parabolic Prandtl-Blasius BL problem.

Region III is the *Blasius BL problem* region governed by the classical Blasius problem, derived below in the framework of low Mach number flow, as a leading-order BL system of ordinary differential equations.

Finally, region IV is the "*non-viscous potential region*" with uniform flow.

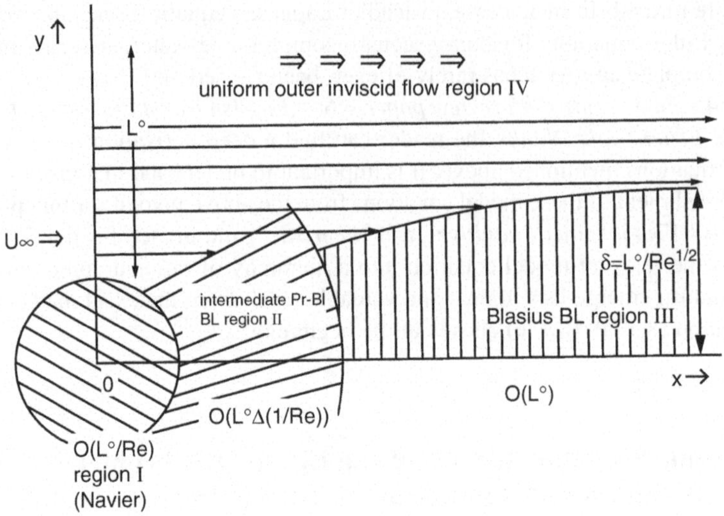


**Fig. 8.9** Sketch of regions I, II, and IV, linked by the Blasius BL region III

### *8.5.1 Formulation of a Steady Compressible 2D Viscous Blasius BL Problem*

In compressible flow, the temperature T has a prescribed value on the flat plate, and if this value is a function of x alone, then the symmetry of the Blasius problem indicates that there should be a 2D flow, with velocity components u, v, and 0, such that u and v are independent of z. The plate is understood to be impermeable, but with a changing temperature, so we impose the following boundary conditions:

$$u = v = 0,\ T = T_0 + \Delta T_0 \Theta(x) \text{ at } y = 0 \text{ for } 0 < x < \infty, \tag{8.31a}$$

where $T_0$ and $\Delta T_0 > 0$ are constant reference temperatures and $\Theta(x)$ is a given function of x. In view of the symmetry, moreover, there is no loss of generality in restricting attention to the half-space $y > 0$.

We start from the complete steady 2D NSF equations for a compressible viscous and heat conducting fluid, with an equation for the temperature assuming that $\mu(T/T_0)$ an $\lambda(T/T_0)$ are the two coefficients of viscosity, and $k(T/T_0)$ the heat conductivity. The above conditions (8.30) and (8.31a) allow us to define a length:

$$l_0 = (\nu_0/U\infty)\,[\Delta T_0/T_0]^2, \text{ where } \nu_0 = \mu(1)/\rho_0, \tag{8.31b}$$

and $\rho_0$ = constant is the "*upstream*" density.

In this case, with a proper choice of non-dimensional quantities, the 2D steady NSF dimensionless equations for a thermally perfect gas, with $p = R\rho T$,

($R = Cp - Cv$, $\gamma = Cp/Cv$), depend in particular on the Reynolds number (Re) and the upstream Mach number ($M\infty$):

$$\mathrm{Re} = l_0 U\infty/\nu_0, \; M\infty = U\infty/(\gamma R T_0)^{1/2}, \tag{8.31c}$$

and in place of the temperature condition in (8.31a), we can write the following dimensionless condition:

$$T = 1 + \tau\Theta(x) \text{ at } y = 0 \text{ for } 0 < x < \infty. \tag{8.31d}$$

From (8.31c) to (8.31d), we write two similarity relations:

$$\tau_0 = (M\infty)^2 \text{ and } \left(1/\mathrm{Re}^{1/2}\right) = (M\infty)^2. \tag{8.31e}$$

In the *dimensionless NSF problem* all the variables are understood in units of their respective scales, where $\Pr = Cp\mu(1)/k(1) \equiv 1$, and if the *Stokes relation* is adopted, then we have $\lambda(1)/\mu(1) = -2/3$.

The paper Godts and Zeytounian, 1991 <7> (see ZAMM, 70 (1), 67–69, 1990) discusses the *dimensionless NSF problem* in the context of the *RAM approach*.

### 8.5.2 Limit Euler (Outer) Equations for $(M\infty)^2 \to 0$, with x and y Fixed

For the *oute*r Euler limit, we write the following asymptotic expansion:

$$\begin{aligned} &u = 1 + (M\infty)^2 u_E + \ldots; \; v = (M\infty)^2 v_E + \ldots, \\ &p = 1 + (M\infty)^4 p_E + \ldots; \rho = 1 + (M\infty)^2 \rho_E + \ldots, \\ &T = 1 + (M\infty)^2 T_E + \ldots, \end{aligned} \tag{8.32a}$$

and derive the following Euler outer model for $\psi_E(x, y)$:

$$\begin{aligned} &u_E = \partial\psi_E/\partial y, \; v_E = -\partial\psi_E/\partial x, \\ &\partial^2\psi_E/\partial x^2 + \partial^2\psi_E/\partial x^2 = 0, \\ &\partial u_E/\partial x + (1/\gamma)\, \partial p_E/\partial x = 0, \\ &p_E(x, y) = -\gamma\partial\psi_E/\partial y, \\ &\psi_E(x, -\infty) = 0. \end{aligned} \tag{8.32b}$$

### 8.5.3 Limit Prandtl (Inner) Equations for $(M\infty)^2 \to 0$, with x and $\eta = y/(M\infty)^2$ Fixed

In this inner representation, with x and $\eta = y/(M\infty)^2$ fixed when $(M\infty)^2 \to \mathbf{0}$, we have the following asymptotic expansion instead of (8.32a):

$$u = u^0{}_{Pr} + (M\infty)^2 u^1{}_{Pr} + \ldots;\ v = (M\infty)^2 v^0{}_{Pr} + (M\infty)^4 v^1{}_{Pr} + \ldots,$$
$$p = 1 + (M\infty)^4 p^1{}_{Pr} + \ldots; \rho = 1 + (M\infty)^2 \rho^1{}_{Pr} + \ldots, \tag{8.33a}$$
$$T = 1 + (M\infty)^2 T^1{}_{Pr} + \ldots,$$

and substitution gives the classical Prandtl BL equations for the incompressible Blasius problem, with $u^0{}_{Pr}$, $v^0{}_{Pr}$ functions of x and η:

$$\partial u^0{}_{Pr}/\partial x + \partial v^0{}_{Pr}/\partial h = O,$$
$$u^0{}_{Pr}\partial u^0{}_{Pr}/\partial x + v^0{}_{Pr}\partial u^0{}_{Pr}/\partial \eta = \partial^2 u^0{}_{Pr}/\partial \eta^2, \tag{8.33b}$$
$$u^0{}_{Pr} = v^0{}_{Pr} = O \text{ at } \eta = 0,\ 0 < x < \infty, \text{and } \lim_{\eta\to\infty} u^0{}_{Pr} = 1.$$

## 8.5.4 *Flow Due to Displacement Thickness*

From (8.33b) for large η (to ∞), we have

$$\lim_{\eta\to\infty} v^0{}_{Pr} = (\beta/2)\, x^{-1/2}, \tag{8.33c}$$

where β = 1. 7208, and as a consequence,

$$v^0{}_{Pr} = -\partial\psi^0{}_{Pr}/\partial x \Rightarrow \beta x^{-1/2},\ \text{for}\,\eta \to \infty. \tag{8.33d}$$

Then *matching* with the outer Euler stream function gives a boundary condition for the above Laplace equation in (8.32b), relative to $\psi_E$, and because the displacement speed vanishes at y = 0, we obtain

$$\partial\psi_E/\partial y = 0 \text{ on } y = 0. \tag{8.33e}$$

## 8.5.5 *Limit BL 2D Steady Equations Due to a Slight Compressibility Effect*

$$\partial u^1{}_{Pr}/\partial x + \partial v^1{}_{Pr}/\partial\eta + u^0{}_{Pr}\partial\rho^1{}_{Pr}/\partial x + v^0{}_{Pr}\partial\rho^1{}_{Pr}/\partial\eta = 0,$$
$$u^0{}_{Pr}\partial u^1{}_{Pr}/\partial x + v^0{}_{Pr}\partial u^1{}_{Pr}/\partial\eta + u^1{}_{Pr}\partial u^0{}_{Pr}/\partial x + v^1{}_{Pr}\partial u^0{}_{Pr}/\partial\eta$$
$$+\rho^1{}_{Pr}\{u^0{}_{Pr}\partial u^0{}_{Pr}/\partial x + v^0{}_{Pr}\partial u^0{}_{Pr}/\partial\eta\} + (1/\gamma)\,\partial p^1{}_{Pr}/\partial x$$
$$= \partial^2 u^1{}_{Pr}/\partial\eta^2 + (d\mu/dT)_{T=1}\left[\partial^2 u^0{}_{Pr}/\partial\eta^2 + (\partial T^1{}_{Pr}/\partial\eta)\,\partial u^0{}_{Pr}/\partial\eta\right], \partial p^1{}_{Pr}/\partial\eta = 0;$$
$$u^0{}_{Pr}\partial T^1{}_{Pr}/\partial x + v^0{}_{Pr}\partial T^1{}_{Pr}/\partial\eta = \partial^2 T^1{}_{Pr}/\partial\eta^2 + (\gamma - 1)\,(\partial u^0{}_{Pr}/\partial\eta)^2, \rho^1{}_{Pr} = -T^1{}_{Pr.} \tag{8.34}$$

Matching with the Euler outer flow gives first

$$p^1{}_{Pr}(x) = 0, \text{ and } u^1{}_{Pr} = 0, \; T^1{}_{Pr} = 0, \text{ when } \eta \to \infty, \tag{8.35a}$$

and the boundary conditions for the above BL equations with the effect of weak compressibility are

$$u^1{}_{Pr} = 0, \; v^1{}_{Pr} = 0, \; T^1{}_{Pr} = \Theta(x) \text{ at } \eta = 0, \text{ for } 0 < x < \infty. \tag{8.35b}$$

*It is clear to me that, without the RAM approach, the derivation of the above BL equations taking into account the complementary terms due to a slight compressibility effect would be quite impossible. The RAM approach thus opens up opportunities for reconsidering the family of incompressible viscous or non-viscous problems, including fluid flow problems with weak compressibility effects!*

From the first equation (of continuity) in (8.34), one can introduce a BL stream function $\psi^1{}_{Pr}(x, \eta)$ such that

$$u^1{}_{Pr} - u^0{}_{Pr}T^1{}_{Pr} = \partial\psi^1{}_{Pr}/\partial\eta \text{ and } v^1{}_{Pr} - v^0{}_{Pr}T^1{}_{Pr} = -\partial\psi^1{}_{Pr}/\partial x. \tag{8.35c}$$

Finally, if we assume $\Theta(x) \equiv 1$ for $0 < x < \infty$ in (8.35b), then a self-similar solution exists on the plate. The reader can find, associated with the system of equations (8.34), the *linear non-homogeneous equation for the function*

$$F(Y) \text{ where } Y = \eta/\sqrt{x}, \tag{8.36}$$

and its numerical treatment by Godts, with some results and concluding remarks, in the paper *Godts and Zeytounian* [2], cited in Chap. 1.

We may make two fundamental points concerning the above asymptotics: first we have the *hyposonic viscous-compressible similarity relation,*

$$\mathrm{Re}\,(M\infty)^4 = O(1), \tag{8.37a}$$

and second, we have the *skin friction coefficient* $C_f$, which contains a new term as a direct consequence of the *slight compressibility effect coupled with a small viscosity effect*:

$$C_f = 0.6641\,(\mathrm{Re}_x)^{-1/2} + (\mathrm{Re}_x)^{-1/2}M_\infty^2\left\{0.6641 + 2F''(0) + 0.6641\,(d\mu/dT)_{T=1}\right\} + \ldots, \tag{8.37b}$$

where $\mathrm{Re}_x = U\infty\,(x/\nu_0)$ is the local Reynolds number, a function of the dimensional distance x from the leading edge of the solid flat plate.

Naturally, one could generalise the above asymptotic analysis to an arbitrary body, with a full numerical computation. Such a generalisation would be interesting for various low-speed motions (up to $\approx$ 300 km/h), as in "road vehicle aerodynamics".

## 8.6 A Theory for Lee Waves Downstream of a Mountain

In the case of lee wave motion downstream of a mountain, an important similarity rule can be written between the Mach number M and the Boussinesq number *Bo*, which is the ratio of the characteristic vertical length scale $H_c$ of the lee wave motion to the height $H_S = RT_c/g$ of a homogeneous hydrostatic atmosphere. In particular, we write

$$Bo/M = B^* = 0(1), \tag{8.38a}$$

where $Bo = H_c/(RT_c/g)$ and $M = U_c/(\gamma RT_c)^{1/2}$, and

$$B^* \approx 1 \Rightarrow H_c \approx (U_c/g)[RT_c/\gamma]^{1/2}, \tag{8.38b}$$

and for typical meteo values we obtain $H_c \Rightarrow H_B \approx 10^3$m, which is obviously a very strong restriction for the application of the Boussinesq equations to atmopheric motions—for the *prediction of lee waves in the whole troposphere* ($\approx$10 km)—around and downstream of a mountain!

But *for me*, in 1967, the first question was *simply the possibility of such an asymptotically consistent derivation of the famous approximate (Boussinesq) equations according to the 1903 Boussinesq assertion: the variations in the density ρ are taken into account only when they modify the action of the gravitational force* $\mathbf{g} = -g\mathbf{k}$, *where* **k** *is the unit vector along the vertical axis z (altitude).*

Indeed, this "*first question*" was posed by *Paul Germain* when the present author was a research engineer in the Aerodynamics Department of ONERA (Chatillon, France), writing his doctoral thesis about theoretical and numerical results on lee waves in the atmosphere, as obtained in the Hydrometeo Center, Moscow, during the period 1961–1966, under the supervision of I A Kibel.

### *8.6.1 2D Steady Model Equations for Lee Waves*

It was not until *1974* that I published the derivation of the *Boussinesq model equations* as reduced equations derived from the full dimensionless NSF system of compressible, viscous, and heat conducting equations, and the reader can find these equations in Chap. 8 of our 1990 monograph [22], cited in "Introduction". But for the lee wave problem, it is usual to derive these *model equations* from the *non-viscous Euler equations*.

### 8.6.2 *Isochoric Case*

As a first *equation à la Long (1953)* for the *variation of the vertical position of a streamline relative to its unperturbed position at upstream infinity* (see Fig. 8.10, reproduced from p. 107 of [13], cited in "Introduction"), we take

$$\begin{aligned} &\Delta(x, z) = z + (1/U\infty)\psi \equiv z - z\infty(\psi), \\ &\psi = \psi(x, z), \\ &u = -\partial\psi/\partial z, \ w = \partial\psi/\partial x, \end{aligned} \tag{8.39a}$$

and obtain from the 2D steady-state Euler equations the following elliptical equation with constant coefficients:

$$\begin{aligned} &\partial^2\Delta/\partial x^2 + \partial^2\Delta/\partial z^2 + [U\infty/N^\circ]\Delta \\ &+ (1/2g)N^{\circ 2}\left\{(\partial\Delta/\partial x)^2 + (\partial\Delta/\partial z)^2 - 2\,(\partial\Delta/\partial z)\right\} = 0. \end{aligned} \tag{8.39b}$$

For *isochoric motion* (density $\rho = R(\psi)$, conservative along the streamlines), in a duct that has a curvilinear bottom—with a "mountain" between the abscissas $x = -l^\circ/2$ and $x = +l^\circ/2$—and an upper flat roof at $z = H\infty$, we can write the following boundary conditions for (8.39b):

$$\begin{aligned} &\Delta(x, h^\circ\eta(x/l^\circ)) = h^\circ\eta(x/l^\circ), \ \ x \in [-l^\circ/2, +l^\circ/2], \\ &\Delta(-\infty, z\infty) = 0, \\ &\Delta(x, H\infty) = 0, \end{aligned} \tag{8.39c}$$

and

$$\Delta(x, z) \text{ is bounded at downstream infinity.} \tag{8.39d}$$

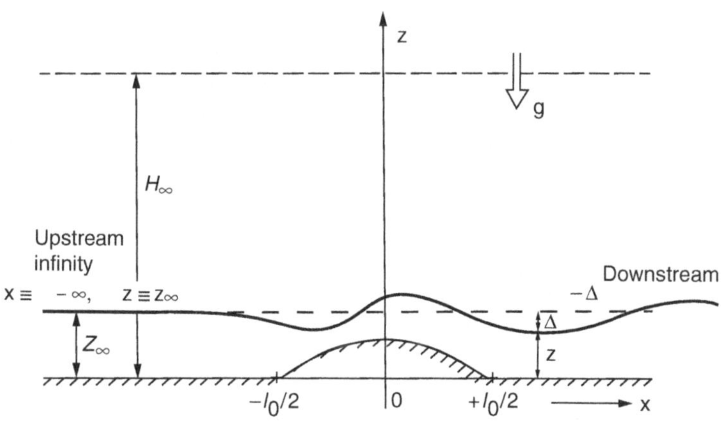


**Fig. 8.10** Lee wave isochoric problem

If we assume a *free surface* at $z = H\infty$, which is a *streamline*, then instead of the last condition (8.39c), we can write

$$(\partial\Delta/\partial x)^2 + (\partial\Delta/\partial z)^2 - 2\,(\partial\Delta/\partial z) + 2\left[g/U\infty^2\right]\Delta = 0, \quad \text{on } z = H\infty + \Delta. \tag{8.40}$$

In equation (8.39b), the parameter $N°$ is the internal frequency of Brunt-Väisälä, which corresponds to influence from the term

$$\text{dlog}\rho_\infty/\text{d}z_\infty = (1/g)N^{\circ 2} = \text{const.}, \tag{8.41}$$

where $\rho_\infty(z_\infty)$ is the density at upstream infinity for $x \to -\infty$.

### 8.6.3 Boussinesq Case

In the *Boussinesq case* with the above similarity relation (8.38a) for the function

$$\psi - z = -\delta(x, z) \text{ and } K_o^2 = B^* \Lambda^\circ/\gamma, \tag{8.42}$$

we derive the following Helmholtz equation, with $K_O^2 = O(1)$:

$$\partial^2\delta/\partial x^2 + \partial^2\delta/\partial z^2 + K_O^2\delta = 0, \tag{8.43a}$$

with

$$\begin{aligned} &z = \nu h(x): \quad \delta = \nu h(x), \\ &z = 1: \quad \delta = 0, \\ &x \to -\infty: \delta \to 0 \text{ and } x \to +\infty: |\delta| < \infty. \end{aligned} \tag{8.43b}$$

From a mathematical point of view, the dominant feature of the above *lee wave problem* (8.43a)–(8.43b) is that the linearity of the Helmholtz equation (8.43a) in the *Boussinesq case is not related to any small perturbation hypothesis*! But one (very) important difficulty remains, concerning the slip boundary condition on the wall of the obstacle (the *mountain* in the case of lee waves): $\delta(x, z) = \nu\, h(x) = 0$, which is a *nonlinear condition* and cannot be linearized without the hypothesis of small disturbances.

The reader can find a detailed resolution of this lee wave problem (8.43a), with (8.43b)—the so-called *Long problem (à la Boussinesq)*—in Sect. *26 of Chap. VI of our 1991 book <3>*, cited in Chap. 2.

### 8.6.4 From the Isochoric Case to the Boussinesq Case

We return to above problem (8.39b)–(8.39c) and introduce the following dimensionless function and variables (marked with stars):

$$\delta^* = \Delta/h^\circ,\ x^* = x/l^\circ,\ z^* = z/H\infty, \tag{8.44a}$$

where $h^\circ$, $l^\circ$, and $H\infty$ are three length scales which are present in the boundary conditions (8.39c). For the dimensionless $\delta^*(x^*, z^*)$, instead of (8.39b) we obtain the following dimensionless equations (dropping the stars):

$$\varepsilon^2\partial^2\delta/\partial x^2 + \partial^2\delta/\partial z^2 + \sigma^2\delta + (1/2)\,\nu\,\mathrm{Bo}\left[\varepsilon^2(\partial\delta/\partial x)^2 + (\partial\delta/\partial z)^2\right]$$
$$- \mathrm{Bo}\,(\partial\delta/\partial z) = 0, \tag{8.44b}$$

with dimensionless parameters $\varepsilon^2$, $\nu$, Bo and $\sigma^2 = O(1)$. When Bo tends to zero, in the *Boussinesq case,* we recover the above problem (8.43a)–(8.43b), and *we can consider this Boussinesq case as a degenerate case of the isochoric one!* For the dimensionless equation (8.44b), we take the dimensionless conditions:

$$\begin{aligned} &\delta(x, \nu h(x)) = h(x),\ x\ [-1/2, +1/2],\\ &\delta(x, 1) = 0,\\ &\delta(-\infty, z\infty) = 0 \quad \text{and} \quad |\delta(+\infty, z)| < \infty. \end{aligned} \tag{8.44c}$$

### 8.6.5 Some Results Concening the Problem (8.44b)–(8.44c)

This problem is complicated, even numerically! Below we give some results performed by *Khiri*, with numerical assistance from Godts. First, one must have a simpler slip condition, and for this we introduce the following new vertical variable:

$$\eta = [z - \nu h(x)]/[1 - \nu h(x)],\ \text{with}\ \phi(x, \eta) = \delta(x, z) + (\eta - 1)h(x). \tag{8.45a}$$

We may then derive a *second-order partial differential equation* with variable coefficients for the function $\phi(x, \eta)$:

$$\varepsilon^2\partial^2\phi/\partial x^2 + A\partial^2\phi/\partial\eta^2 + 2\,\mathrm{Bo}\partial^2\phi/\partial x\,\partial\eta + \varepsilon^2(\mathrm{Bo}/2)\nu(\partial\phi/\partial x)^2$$
$$+ C\,(\partial\phi/\partial x)\,\partial\phi/\partial\eta + D(\partial\phi/\partial\eta)^2 + E\,(\partial\phi/\partial\eta) - F(\partial\phi/\partial x)$$
$$+ \sigma^2\phi = H(x, h(x)), \tag{8.45b}$$

with conditions

$$\phi(x,0) = 0\,,\ \phi(x,1) = 0,\ \ \phi(-\infty,\eta) = 0,\ \phi(+\infty,\eta)\ \text{is bounded.} \tag{8.45c}$$

In equation (8.45b), the coefficients A, B, C, D, E, F and the function H are given in *Khiri's* doctoral thesis (University of Lille I, 1992).

Below we consider three different approximate models for the various values of $\sigma^2$.

*Linear Case* when $l^\circ \gg h^\circ$, $v \to 0$, $\varepsilon^2$ and Bo are fixed, and we obtain a linear equation:

$$\varepsilon^2 \partial^2\phi/\partial x^2 + \partial^2\phi/\partial\eta^2 - \mathrm{Bo}\,\partial\phi/\partial\eta + \sigma^2\phi = H_L(x), \tag{8.45d}$$

where

$$H_L(x) = (\eta - 1)\left[\varepsilon^2 d^2h(x)/dx^2 + \sigma^2 h(x)\right] - \mathrm{Bo}\,h(x). \tag{8.45e}$$

This linear case is significant only for a weakly sloping ($l^\circ \gg h^\circ$) curvilinear bottom.

*Long Wave Case* when $H\infty \ll l^\circ$, $\varepsilon^2 \to 0$, Bo and $v$ fixed. The associated equation is

$$\begin{aligned}&\partial^2\phi/\partial\eta^2 + v\,(\mathrm{Bo}/2)\,(\partial\phi/\partial\eta)^2 - \mathrm{Bo}\,\partial\phi/\partial\eta +\\ &+\sigma^2[1 - vh(x)]^2\phi = -h(x)\{\sigma^2(1-\eta)\\ &[1 - vh(x)]^2 + \mathrm{Bo}[1 - (v/2)h(x)]\},\end{aligned} \tag{8.46a}$$

where x plays the role of a parameter via h(x).

*Boussinesq Case* when $(U\infty)^2/g \ll H\infty \ll g/N^{\circ 2}$, $\mathrm{Bo} \to 0$, $\varepsilon^2$, $\sigma^2$ and $v$ fixed. In this case, we obtain once again a linear (Boussinesq) equation instead of (8.45b):

$$\begin{aligned}&\varepsilon^2\partial^2\phi/\partial x^2 + A\,\partial^2\phi/\partial\eta^2 + 2B\,\partial\phi/\partial x\,\partial\eta + E\,(\partial\phi/\partial\eta) + \sigma^2\phi\\ &\quad = H_B(x),\end{aligned} \tag{8.46b}$$

where A, B, E and also $H_B(x)$ are explicit functions of h(x) and dh/dx, $d^2h/dx^2$.

For the validity of equation (8.46b), $H\infty$ must be appreciably less than the scale related to the density variation far upstream:

$$\rho_\infty(z_\infty) = \rho_\infty(o)\exp\left[-\frac{1}{g}N^{o^2}Z_\infty\right] \tag{8.46c}$$

*Lee Wave Configurations* The solution of the problem (8.45b) with (8.45c) is

$$\phi(x,\eta) = \Sigma_n\phi_n(x)\ \sin(n\pi\eta),\ n = 1, 2, 3, \ldots, \infty, \tag{8.47a}$$

and according to the classical Galerkine method, instead of (8.45b) with (8.45c), we derive a system of *ordinary second-order differential equations* for the "*amplitude*" $\phi_n(x)$, n = 1,2,3,.... In Khiri's thesis, the reader will find this system—*although the reader could also derive it himself*—taking into account the two conditions in x:

$$\phi_n(-\infty) = 0 \text{ and } |\phi_n(+\infty)| \text{ is bounded.} \tag{8.47b}$$

In particular, for n = 1, the *linear case*, we obtain the following simple equation:

$$d^2\phi_1(x)/dx^2 + (1/\varepsilon^2)\left[\sigma^2 - \pi^2\right]\phi_1 = (2/\pi)\left[d^2h/dx^2 + (1/\varepsilon^2)\left(\sigma^2 + 2Bo\right) h(x)\right]. \tag{8.47c}$$

For *long waves, when* $\varepsilon^2$ tends to zero, we derive a simple algebraic equation for $\phi_1(x)$:

$$\phi_1^2 + (3/2\nu\,\pi Bo)\left[\sigma^2(1 - \nu h(x))^2 - \pi^2\right]\phi_1 = H(x), \tag{8.47d}$$

where

$$H(x) = (3/\nu\,\pi^2 Bo)h(x)\left\{\sigma^2(1 - \nu h(x))^2 + 2Bo\,[1 - (\nu/2)h(x)]\right\}. \tag{8.47e}$$

The solution of (8.47d) is *very useful* and can be used, for example, as a *test solution for checking a routine in the numerical integration* of nonlinear equations for the various $\phi_n(x)$.

In Figs. 8.11, 8.12 and 8.13 (reproduced from pp. 164–166 of [13], cited in "Introduction") the reader can find various results relating to the solution of the nonlinear equation (8.47g) with the two conditions (8.47b).

The figures show the *configuration of streamlines*, and we note the nonlinear equation (8.47g) for the function $\phi_1$ (n = 1):

$$\delta(x,\ z) = (1 - \eta)h(x) + \phi_1(x)\ \sin(\pi\eta); \tag{8.47f}$$

and

$$\varepsilon^2 d^2\phi_1/dx^2 + A_1\,(d\phi_1/dx)^2 + B_1\,\phi_1\,(d\phi_1/dx) + C_1\,\phi_1^2 + D_1\,(\partial\phi_1/\partial x) = H_1(x), \tag{8.47g}$$

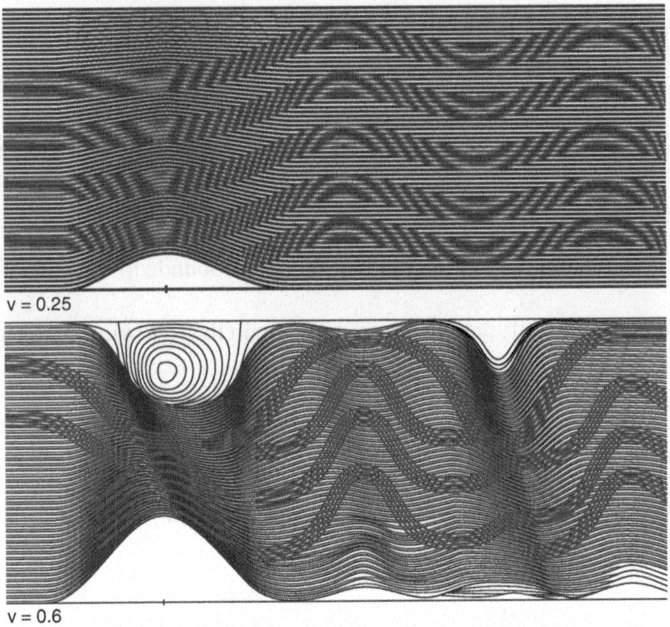


**Fig. 8.11** $\sigma = 2\pi$, $\varepsilon = B_0 = 1$, for $\nu = 0.25$ and $\nu = 0.6$

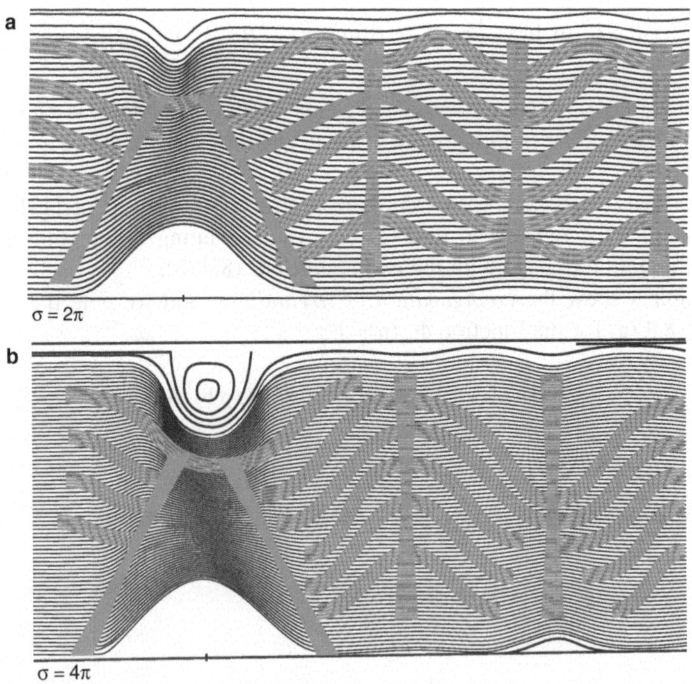


**Fig. 8.12** (**a**) $\nu = 0.5$, $\varepsilon = B_0 = 1$. (**b**) $\nu = 0.5$, $\varepsilon = B_0 = 1$

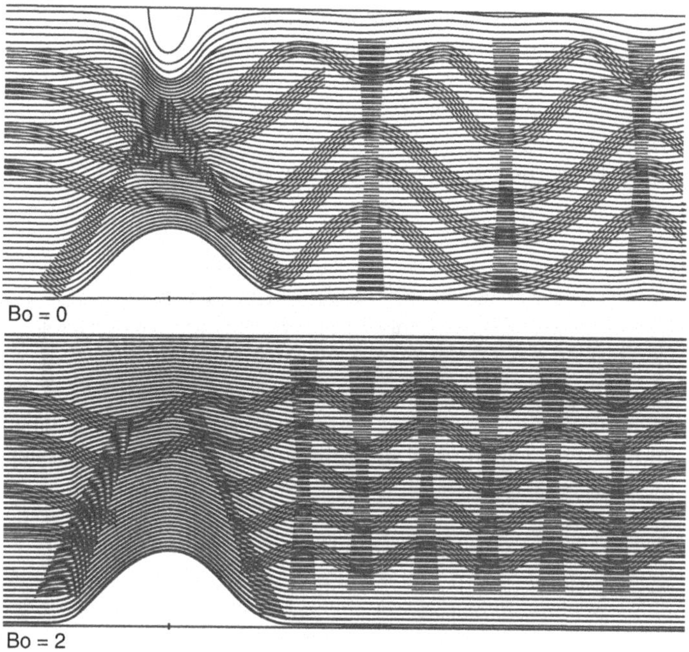


**Fig. 8.13** $\nu = 0.5, \sigma = 2\pi, \varepsilon = 1$

### *8.6.6 Models for Steady 2D Non-viscous Lee Waves in a Baroclinic Compressible Troposphere*

In dimensional form, for a perfect gas, the steady two-dimensional Euler equations for the velocity components u and w, the pressure p, and the density ρ, all assumed to depend only on the coordinates x and z, can be written in the following form:

$$\begin{aligned} &\rho\,[u\,\partial u/\partial x + w\,\partial u/\partial z] + \partial p/\partial x = 0, \\ &\rho\,[u\,\partial w/\partial x + w\,\partial w/\partial z] + \partial p/\partial z + g\rho = 0, \\ &\partial(\rho u)/\partial x + \partial(\rho w)/\partial z = 0, \\ &[u\,\partial/\partial x + w\,\partial/\partial z](p/\rho^{\gamma}) = 0, \\ &p = R\rho\,T. \end{aligned} \tag{8.48a}$$

We consider a mountain, and write the equation in the plane of the gravitational field as

$$z = h^0\eta(x/l_0), \quad -12 \le x/l_0 \le +l2, \tag{8.48b}$$

while the slip condition gives

$$w = (d\eta/dx)\,u \quad \text{on } z = h^0\eta(x/l_0) \tag{8.48c}$$

Again, our non-viscous, non-heat-conducting fluid is a perfect gas, and by introducing a stream function ψ(x,z) such that

$$\rho u = -\partial\psi/\partial z \text{ and } \rho w = +\partial\psi/\partial x, \tag{8.48d}$$

and using (8.48d), we derive, instead of the system (8.48a), the following three equations:

$$\begin{aligned} &p = \rho^{\gamma}\Pi(\psi)\,, \\ &1/2\,(u^2 + w^2) + [\gamma/\,(\gamma - 1)]\,\rho^{\gamma-1}\Pi(\psi) + gz = I(\psi), \\ &\omega \equiv \partial u\,\partial z - \partial w\,\partial x = -\rho\{dI/d\psi - (1/(\gamma - 1)\,[p/\rho]\,d\log\Pi/d\psi\}, \end{aligned} \tag{8.48e}$$

where Π(ψ) and I(ψ) are conservative along each streamline, and are determined from the behaviour in the unperturbed upstream region (with $x \to -\infty$) and where $z_\infty$ is the altitude of the unperturbed streamline. We assume that in this unperturbed upstream region

$$\begin{aligned} &u = U_\infty(z_\infty) \text{ and } w = 0, \\ &p = p_\infty(z_\infty) \text{ and } \rho = \rho_\infty(z_\infty)\,, \\ &T = T_\infty(z_\infty)\,, \end{aligned} \tag{8.48f}$$

at $x \to -\infty$. Then with (8.48f), if we take into account

$$T = T_\infty - [(\gamma - 1)/\,\gamma R]\{(1/2)\,(u^2 + w^2) - U^2{}_\infty(z_\infty)] + g(z - z_\infty)\}, \tag{8.48g}$$

for the density ρ, instead, we obtain the relation :

$$\rho = \rho_\infty\{1 + [(T - T_\infty)/T_\infty]\}^{1/(\gamma-1)}. \tag{8.48h}$$

Finally, from the above results, we obtain an equation for ψ(x,z) which has a very "*awkward*" form. However, we shall not write this equation here.

For present purposes, we introduce (instead of ψ) the vertical deviation δ(x, z) of a streamline in the *perturbed flow over a mountain* relative to the *unperturbed* flow at infinity upstream. Hence, we write

$$\begin{aligned} &z = z_\infty(\psi) + \delta(x,\,z), \\ &\partial\psi\,\partial x = -\rho_\infty U_\infty \partial\delta/\partial x, \\ &\partial\psi\,\partial z = \rho_\infty U_\infty(1 - \partial\delta/\partial z), \end{aligned} \tag{8.48i}$$

and in place of the slip condition(8.48c), we obtain

$$\delta(x, h^0\eta(x/l_0)) = h^0\eta(x/l_0). \tag{8.48j}$$

For the function δ(x, z), we derive the following second-order partial differential equation (see pp. 256–262 in our LNP 672, 2006):

$$
\begin{aligned}
&\partial^2\delta/\partial x^2 + \partial^2\delta/\partial z^2 + \left[g/(U_\infty)^2\right] (\rho/\rho_\infty)^2 N^2(z_\infty)\,\delta \\
&\quad = -(1/2)\,(\rho/\rho_\infty)^2 d/dz_\infty\left\{\log\left[(U_\infty)^2 \exp(-\sigma_\infty/Cp)\right]\right\} \\
&\quad + (½)\, d/dz_\infty\left\{\log\left[(\rho_\infty)^2(U_\infty)^2\exp(-\sigma_\infty/Cp)\right]\right\}\left[(\partial\delta/\partial x)^2\right. \\
&\quad \left. + (\partial\delta/\partial z)^2 + 1 - 2\,(\partial\delta/\partial z)\right] + (\partial\log\rho/\partial x)(\partial\delta/\partial x) \\
&\quad + (\partial\log\rho/\partial z)\left(\frac{\partial\delta}{\partial z} - 1\right).
\end{aligned}
\tag{8.49a}
$$

where $\sigma_\infty \equiv C_v \log\left(-P_\infty/\rho_\infty^\gamma\right)$.

The derived above Eq. (8.49a) is an accurate consequence of the Euler system of Eq. (8.48a), and for this Eq. (8.49a), with (8.48j), we have also the following three conditions:

$$
\begin{aligned}
&\delta(x = -\infty, z_\infty) = o, \\
&\delta(x, z = H_\infty) = o, \\
&|\delta(x = +\infty, z)| < \infty,
\end{aligned}
\tag{8.49b}
$$

where $H_\infty$ is the altitude of the upper level (for instance, the altitude of the tropopause, assumed to be a flat, horizontal plane), where the streamlines are *undeflected*. The last condition of (8.49b) is the *only possible physical one, because of the lee-waves regime downstream of the mountain*.

However, in Eq. (8.49a) for the function δ(x, z), we have also, as an second *unknown function*, the *density* ρ, and consequently we must return to the two above relations, (6.48g, 6.48h), which we transform to an equation for ρ in which δ is present:

$$
\begin{aligned}
(\rho/\rho_\infty)^{\gamma-1} = &\left\{1 + \left[(U_\infty)^2/2CpT_\infty\right](\rho_\infty/\rho)^2\left[(\partial\delta/\partial x)^2 + (\partial\delta/\partial z)^2\right.\right. \\
&\left.\left. -2(\partial\delta/\partial z) + 1\right] + \left[(U_\infty)^2/2CpT_\infty\right]\left[\left(2g/(U_\infty)^2\right)\delta - 1\right]\right\},
\end{aligned}
\tag{8.49c}
$$

and we have the following condition at infinity upstream:

$$
\rho \to \rho_\infty(z_\infty) \text{ when } x \to -\infty, \text{ and } \delta(x = -\infty, z_\infty) = o. \tag{8.49d}
$$

The above problem (8.49a)–(8.49d), for the two functions δ and ρ is *strongly nonlinear*. Below, we consider a simplified case when:

$$
U_\infty \equiv U^o{}_\infty = \text{const and } -dT_\infty/dz_\infty \equiv \Gamma^o{}_\infty = \text{const}, \tag{8.50a}
$$

such that

$$
T_\infty(z_\infty) = T_\infty(o)\left[1 - (\Gamma^o{}_\infty/T_\infty(o))z_\infty\right], \tag{8.50b}
$$

and this lineal disribution for $T_\infty(z_\infty)$ is very well justified in the usual meteorological situation in the troposphere, when the lee-waves regime is considered—$H_\infty$ being the height of the whole troposphere, The parameter:

$$\mu^o{}_\infty = H_\infty \Gamma^o{}_\infty / T_\infty(o), \tag{8.50c}$$

is a reference parameter for the temperature profile at infinity upstream, and

$$T_\infty(z_\infty)/T_\infty(o) \equiv \Theta(\zeta_\infty) = 1 - \mu^o{}_\infty \zeta_\infty, \text{ with } \zeta_\infty \equiv z_\infty / H_\infty. \tag{8.50d}$$

**Dimensionless Problem**

Now we introduce the *non-dimensional density perturbation*

$$\varpi = (\rho - \rho_\infty)/\rho_\infty, \tag{8.51a}$$

and also

$$\Delta = \delta / h^o \tag{8.51b}$$

the non-dimensional vertical displacement of the streamline. Since far ahead of the mountain there is a uniform flow with velocity components ($U^o{}_\infty$ = const, o), we write, in place of $z = z_\infty(\psi) + \delta(x, z)$, the following dimensionless relation:

$$\zeta_\infty = \mathrm{Bo}[\zeta - (1/\nu_o)\Delta], \tag{8.51c}$$

where z is reduced by the vertical length scale $H^o$ characterizing the lee waves process and $z_\infty$, is reduced by $H_\infty \equiv RT_\infty(o)/g$. In this case we have two ratios,

$$\nu_o = H^o / h^o, \mathrm{Bo} = H^o / H_\infty, \tag{8.51d}$$

where $\nu_o$ is, indeed, the *'linearization'* parameter, and in place of (8.48j) we have the following dimensionless slip condition:

$$\Delta(\xi, \zeta = (1/\nu_o)\eta(\xi)) = \eta(\xi), \text{ where } \xi = x/l_o. \tag{8.51e}$$

The second condition of (8.49b) gives

$$\Delta(\xi, \zeta = 1/\mathrm{Bo}) = o, \tag{8.51f}$$

and the *'long-wave' approximation parameter* is

$$\varepsilon = H^o / l_o. \tag{8.51g}$$

For $\varpi$, defined above by (8.51a), we obtain the following dimensionless relation from (8.49c): $(1 + \varpi)^{\gamma - 1}$

$$\begin{aligned} &= 1 - [1/\Theta(\zeta, \Delta)]\Big\{(1/2)(\gamma - 1)\big[(M_\infty/\nu_o\big]^2\big[1/(1 + \varpi)^2\big]\big[\varepsilon^2(\partial\Delta/\partial\xi)^2 \\ &+(\partial\Delta/\partial\zeta)^2 - 2\nu_o(\partial\Delta/\partial\zeta) + \nu_o{}^2\big] + \\ &[(\gamma - 1)\mathrm{Bo}/\gamma\nu_o]\Delta - (1/2)(\gamma - 1)(M_\infty)^2\Big\}, \end{aligned} \tag{8.52a}$$

where

$$\Theta(\zeta, \Delta) = 1 - \mathrm{Bo}\,\mu^{\mathrm{o}}{}_{\infty}[\zeta - (1/\nu_{\mathrm{o}})\Delta]. \tag{8.52b}$$

Finally, for the function Δ, we obtain the following *dimensionless equation* from (8.49a), when we take into account (8.50a–8.50d):

$$\begin{aligned}&\Theta(\zeta, \Delta)\Big\{\varepsilon^2(\partial\Delta/\partial\xi)^2 + (\partial\Delta/\partial\zeta)^2 - [1/(1+\varpi)]\,[\varepsilon^2(\partial\Delta/\partial\xi)(\partial\varpi/\partial\xi)\\&+(\partial\Delta/\partial\zeta)(\partial\varpi/\partial\xi) - \nu_{\mathrm{o}}(\partial\varpi/\partial\zeta)\} + [\mathrm{Bo}/\gamma\mathrm{M}_\infty]^2 S_{\mathrm{o}}(1+\varpi)^2\Delta\\&= (\nu_{\mathrm{o}}/2)\mathrm{Bo}\,S_{\mathrm{o}}\Big\{\varpi(2+\varpi) - \Big[\varepsilon^2(\partial\Delta/\partial\xi)^2 + (\partial\Delta/\partial\zeta)^2 - 2\nu_{\mathrm{o}}(\partial\Delta/\partial\zeta)\Big]\Big\},\end{aligned} \tag{8.52c}$$

where the "*hydrostatic stability parameter*" is

$$S_o[(\gamma-1)/\gamma] - \mu^{\mathrm{o}}{}_{\infty}, \tag{8.52d}$$

characterizing the *stratification* of the *unperturbed flow at infinity upstream.*

For typical meteorological values in the troposphere for dry air, we have

$$\begin{aligned}&\gamma = 1.4, \mathrm{T}_\infty(0) = 288\,^\circ\mathrm{C},\\&\mu^{\mathrm{o}}{}_{\infty} = (\mathrm{R}/\mathrm{g})\Gamma^{\mathrm{o}}{}_{\infty} \approx 0.19037,\\&[(\gamma-1)/\gamma] - \mu^{\mathrm{o}}{}_{\infty} \approx 0.09534.\end{aligned} \tag{8.52e}$$

On the other hand, $[\gamma \mathrm{RT}_\infty(0)]^{1/2} \approx 340.17$ m/s, and if 34 m/s $\geq \mathrm{U}^{\mathrm{o}}{}_{\infty} \geq$ 10 m/s, then we obtain $0.03 \leq \mathrm{M}_\infty \leq 0.1$.

We also observe that, for the unknown function $\Delta(\xi, \zeta)$ solution of the above Eq. (8.52c) we have the following boundary conditions for the lee wave problem over and downstream of the mountain:

$$\Delta(\xi, (1/\nu_0)\eta(\xi)) = \eta(\xi), \; -1/2 \leq \xi \leq +1/2, \tag{8.52f}$$

$$\Delta(\xi = -\infty, \zeta_\infty) = 0,\; \Delta(\xi, \zeta = (1/\mathrm{Bo})) = 0,\; |\Delta(\xi = +\infty, \zeta)| < \infty. \tag{8.52g}$$

In equation (8.52c), the parameter

$$\mathrm{Ko}^2 \equiv (S_0/\gamma)[\mathrm{Bo}/\mathrm{M}_\infty]^2 = g[(\Gamma_{\mathrm{A}} - \Gamma^{\mathrm{o}}{}_{\infty})/\mathrm{T}_\infty(0)]\,(\mathrm{H}^0/\mathrm{U}^{\circ}{}_{\infty}), \tag{8.53a}$$

where $\Gamma_{\mathrm{A}} = (g/\mathrm{R})[(\gamma - 1)/\gamma]$ is the dry adiabatic temperature gradient. It plays a *fundamental* role when

$$\mathrm{M}_\infty = \mathrm{U}^{\circ}{}_{\infty}/\,[\gamma \mathrm{RT}_\infty(0)]^{1/2} << 1, \;\text{ but } \mathrm{Ko}^2 = \mathrm{O}(1). \tag{8.53b}$$

For the *lee wave problem*, it is essential that

$$\mathrm{Ko}^2 = \mathrm{O}(1), \text{ but not } << 1 \text{ or } >> 1! \tag{8.53c}$$

### 8.6.7 Four Limiting Cases

We consider the following four cases:

(i) The first case relates to *deep convection* and is valid in the *whole troposphere* when $\mathrm{Bo} = \mathrm{O}(1)$, but with the following similarity rule:

$$S_0 = \mathrm{S}^* \,(\mathrm{M}_\infty)^2 \Leftrightarrow \; \mathrm{Ko}^2 = (\mathrm{S}^*/\gamma)\mathrm{Bo}^2, \; \mathrm{S}^* = \mathrm{O}(1). \tag{8.54a}$$

(ii) The second case is the *Boussinesq case,* when the *Boussinesq* and *Mach* numbers *are both small parameters*, i.e., $\mathrm{Bo} \ll 1$ and $\mathrm{M}\infty \ll 1$, but when $S_0 = \mathrm{O}(1)$, with a similarity relation between Bo and $\mathrm{M}_\infty$ :

$$\mathrm{Bo}/\mathrm{M}_\infty = \mathrm{B}^* = \mathrm{O}(1) \; \Leftrightarrow \mathrm{H}^0 \approx \; (\mathrm{U}^\circ{}_\infty/g)[\mathrm{RT}_\infty(0)/\gamma]^{1/2}. \tag{8.54b}$$

(iii) The third case is the isochoric case, when $\gamma \gg 1$, and the Mach number is low, i.e., $\mathrm{M}_\infty \ll 1$ such that $S_0 = \mathrm{O}(1)$ and $\mathrm{Bo} = \mathrm{O}(1)$, but where

$$\gamma \mathrm{M}_\infty = \mathrm{M}^* = \mathrm{O}(1) \; \Leftrightarrow \; \mathrm{U}^\circ{}_\infty \approx \; [\mathrm{RT}_\infty(0)]^{1/2}. \tag{8.54c}$$

(iv) The fourth case relates to a very thin atmospheric layer, when $S_0$ $\mathrm{Bo} \approx 1$ with $\mathrm{S}_0 \gg 1$, $\mathrm{Bo} \ll 1$, and $\gamma\,(\mathrm{M}_\infty)^2 \ll 1$, such that

$$\mathrm{Bo}/\gamma\,(\mathrm{M}_\infty)^2 \approx \; 1 \; \Leftrightarrow \mathrm{H}^0 \approx \; (\mathrm{U}^\circ{}_\infty)^2/g, \text{and } (\Gamma_\mathrm{A} - \Gamma^\circ{}_\infty) \\ = (\mathrm{U}^\circ{}_\infty)^2/g\mathrm{T}_\infty(0). \tag{8.54d}$$

For each of the above cases, we can derive a significant low-Mach-number reduced model problem from the full nonlinear problem (8.52c) with (8.52d)–(8.52f) for the function $\Delta(\xi,\zeta)$, together with the relations (8.52a,b).

We observe that in case (i), if we also assume that $\nu_0 \gg 1$ or, according to the first relation of (8.51c), that

$$\mathrm{h}^0 << \mathrm{H}^0 \; \text{ and } \; \nu_0(\mathrm{M}_\infty)^2 = \mathrm{O}(1), \tag{8.55}$$

we can derive from (8.52c), as *leading-order equation,* when $\mathrm{M}_\infty \to 0$, a reduced equation for $\Delta_0(\xi, \zeta)$, which is very similar to an equation *derived rather in an "ad hoc" manner* and considered by *Dorodnitsyn* (see pp. 324–328 in our monograph (1990) [22], cited in "Introduction").

## 8.7 A Model Problem for a Local Thermal Spot Effect

Here the considered problem is *two-dimensional* (x, z) and *steady* $\partial/\partial t = 0$, and we write *the thermal boundary condition* in the form

$$T/T_{\infty}(0) = 1 + \tau_0 \Theta(x/l_0), \quad \text{on } z = 0, \tag{8.56}$$

where (the local thermal spot) $\Theta(x/l_0) \equiv 0$, if $|x/l_0| \leq 1$.

Far upstream, when $x \rightarrow \infty$ and $\Theta(x/l_0) \equiv 0$, we assume that we have a basic undisturbed flow which is characterized by an *Ekman layer* profile. Then, in dimensionless form,

$$U_{Ek}(x/l_0, (z/l_0)/\kappa_0) = U_g(x/l_0)\{1 - \exp[-(z/l_0)/\kappa_0] \cos((z/l_0)/\kappa_0)\}, \tag{8.57}$$

where

$$\begin{aligned} \kappa_0 &= [\Omega_0 \sin\varphi_0/v_0]/l_0 \equiv [Re_{l0}/2Ro_{l0}]^{-1/2}, \\ Re_{l0} &= l_0 U_0/\nu_0, \\ Ro_{l0} &= U_0/2 l_0 \Omega_0 \sin\varphi_0, \end{aligned} \tag{8.58}$$

with the *ratio* ($\kappa_0$), <of> *local Reynolds* ($Re_{l0}$) *and* ($2Ro_{l0}$) *2Rossby numbers*, based on the local, scale length $l_0$.

### 8.7.1 Dimensionless Local Problem

If we nondimensionalize the coordinates x and z with $l_0$, then we introduce in the dimensionless problem the following *(local) Boussinesq* number:

$$Bo_{l0} = g\, l_0/R\, T_{\infty}(0). \tag{8.59a}$$

If $l_0 \sim 10^3$ m, then $Bo_{l0} \ll 1$, but if $l_0 \sim 10^3$ m, then we have first $2Ro_{l0} \gg 1$ and also $2Ro_{l0} \ll Re_{l0}$. Therefore, in this case (when $l_0 \sim 10^3$ m), we can assume that

$$\begin{aligned} 2Ro_{l0} &= Re_{l0}^{-1/a}, \\ \kappa_0 &= Re_{l0}^{-1/m}, \\ m &= 2a/(a-1) > 2, \; a \neq 1, \end{aligned} \tag{8.59b}$$

where the scalar $a > 0$ is chosen below. The reader can check that for $U_0 \sim 10$ m/s, $\nu_0 = 5$ m$^2$/s, and $f_0 \equiv 2\Omega_0 \sin\varphi_0 = 10^{-4}$ 1/s, the case of $l_0 \sim 10^3$ m leads to $m = 5$. For this case we can use

$$l_0 \sim (U_0/g)[RT_\infty(0)/\gamma]^{1/2} \Rightarrow Bo_{l0}/M \equiv B^* = 1, \tag{8.59c}$$

where $M = U_0/[\gamma\, RT_\infty(0)]^{1/2}$, and the Boussinesq approximation is correct.

The value $m = 5$ is the same as the one used by Smith and Smith et al. for the flow over a short isolated two-dimensional hump in the boundary layers. For the stratified Boussinesq fluid, see the work of Sykes (1978). When $m = 5$, and (the below relation (8.59d) being a *constraint* for $L_0$ !)

$$l_0/L_0 \sim Re_{L0}^{-3/8}, \text{ where } Re_{L0} = U_0 L_0/\nu_0, \tag{8.59d}$$

we recover the famous "3/8"! With $m = 5$ we obtain $a = 5/3$.

According to the *Boussinesq approximation* and taking into account the above relations, we can formulate the following dimensionless local problem, if we impose

$$\tau_0 \to 0, \text{ with } M \to 0, \text{and } \tau_0/M = \lambda_0 \approx 1, \tag{8.59e}$$

$$\begin{aligned}
&u\partial u/\partial x + w\partial u/\partial z + (1/\gamma)\partial\pi/\partial x = \kappa_0{}^2\left[\partial^2 u/\partial x^2 + \partial^2 u/\partial z^2\right],\\
&u\partial w/\partial x + w\partial w/\partial z + (1/\gamma)\partial\pi/\partial z - (B^*/\gamma)\theta = \kappa_0{}^2\left[\partial^2 w/\partial x^2 + \partial^2 w/\partial z^2\right],\\
&\partial u/\partial x + \partial w/\partial z = 0\,.
\end{aligned} \tag{8.60a}$$

$$\begin{aligned}
&u\partial\theta/\partial x + w\partial\theta/\partial z + B^*\left[(\gamma-1)/\gamma + (dT_\infty/dz_\infty)_{z\infty=0}\right] w\\
&= (1/Pr)\kappa_0{}^5\left[\partial^2\theta/\partial x^2 + \partial^2\theta/\partial z^2\right],\\
&\omega = -\theta,\\
&z = 0: u = w = 0,\ \theta = \lambda_0\Theta(x),\ 0 < x < 1,\\
&x \to -\infty: u \to 1 - \exp(-z/\kappa_0)\cos(z/\kappa_0) \equiv U^\infty(z/\kappa_0),\\
&\theta = \pi = -\omega \to 0.
\end{aligned} \tag{8.60b}$$

We observe that, in (8.56), for $\tau_0 = \Delta T_0/T_\infty(0)$, we have

$$\begin{aligned}
&\Delta T_0 \sim U_0[T_\infty(0)/\gamma R]^{1/2},\\
&U_g(x/l_0) \sim 1 + O(\kappa_0{}^3),\\
&\text{when } k_0 \to 0,
\end{aligned} \tag{8.61a}$$

and we note that if $z/\kappa_0 = z^* \to \infty$, then

$$u \to 1, \text{ for } x \to -\infty, \tag{8.61b}$$

and if $z^* \to 0$, then

$$u \sim z^*, \text{ for } x \to -\infty. \tag{8.61c}$$

But, now if we wish to take into account the *boundary conditions on the ground* $z=0$, we must introduce the inner variable

$$\zeta = z/(\kappa_0)^{\beta}, \beta > 1, \tag{8.61d}$$

and in this case

$$u \sim (\kappa_0)^{\beta-1}\zeta, \text{ for } x \to -\infty. \tag{8.61e}$$

From the first equation in (8.60), with the above relations, we find that, if

$$u \sim (\kappa_0)^{\beta-1} u_l(x, \zeta), \tag{8.61f}$$

then we must impose

$$\beta - 1 = 5 - 2\beta \Rightarrow \beta = 2. \tag{8.62}$$

### 8.7.2 *Triple Deck Structure*

Finally, we establish that *three vertical variables* z, z* and $\zeta$ are necessary for the asymptotic analysis of the above system (8.60):

I. z, for the *upper region*, where

$$u \sim u_{\text{up}} \to 1, \text{when } x \to -\infty. \tag{8.63a}$$

II. $z^* = z/\kappa_0$, for the *middle region*, where

$$u \sim u_{\text{mid}} \to 1 \exp(-z^*) \cos z^* \equiv U^{\infty}(z^*), \text{ when } x \to -\infty. \tag{8.63b}$$

III. $\zeta = z/\kappa_0^2$, for the *lower viscous region close to the wall*, where

$$u \sim \kappa_0\, u_{\text{low}}, \text{ and } u_{\text{low}} \to \zeta, \text{ when } x \to -\infty. \tag{8.63c}$$

For the other case, when $l_0 < (\kappa_0)^3 L_0$ and $l_0 > (\kappa_0)^3 L_0$, one must apply a different asymptotic analysis [see, for example, the work of Smith et al.]. But the cases $m=6$ and $m=4$ can be analysed from the above problem (8.60). For the case $m=3$, one must start from another problem, where the *Boussinesq approximation does not emerge,* and then we have $l_0 \sim 10^4$ m and we may *neglect the Coriolis terms in the local non-Boussinesq equations.*

The reader can find a detailed account of various facets of singular coupling and the triple-deck structure and modelling on pp. 471–525 of our book 2002 [7], cited in "Introduction", FMIA 64, with references on pp. 542–545.

### 8.7.3 Analysis of the Three Regions

We shall give the analysis for the three regions of the triple-deck theory: *upper*, *middle*, and *lower*, with a little discussion. *The above local problem is a very good example for developing a deep understanding of internal asymptotic analysis!*

Beginning in the *middle deck*, where x and $z^* = z/\kappa_0$ are the order one coordinates, we expand the flow variables as below:

$$\begin{aligned} u &= U^\infty(z^*) + \kappa_0{}^a u_{mid} + \dots \\ w &= \quad + \kappa_0{}^b w_{mid} + \dots \\ \pi &= \quad + \kappa_0{}^2 \pi_{mid} + \dots \\ \theta &= \quad + \kappa_0{}^d\, \theta_{mid} + \dots \end{aligned} \tag{8.64a}$$

and substitute in (8.60) to find, for the first lowest order,

$$\begin{aligned} &U^\infty(z^*)\partial u_{mid}/\partial x + (dU^\infty/dz^*)w_{mid} = 0, \\ &\partial u_{mid}/\partial x + \partial w_{mid}/\partial z^* = 0, \\ &\partial \pi_{mid}/\partial z^* = B^*\,\theta_{mid}, \\ &\partial \theta_{mid}/\partial x = 0, \\ &\text{if : } a = 1,\ b = 1 + a, \text{ and } d = 1. \end{aligned} \tag{8.64b}$$

This choice in (8.64b) is necessary if we want to obtain a significant degeneracy of the problem (8.60) in the lower viscous region, in the vicinity of the wall $\zeta = 0$. Notice that the effects of the expansion of the boundary layer are $O(\kappa_0)$ in u and for w, $O(\kappa_0{}^2)$. Furthermore, in the boundary layer, if we take into account the fact that $\beta = 2$, we necessarily have $\pi = \kappa_0{}^2\ \pi_{low}$ and, by continuity, we obtain the form of the expansion for $\pi$ in (8.64a).

Solutions for $u_{mid}(x, z^*)$ and $w_{mid}(x, z^*)$ satisfying the upstream boundary conditions are

$$u_{mid} = A(x)dU^\infty(z^*)/dz^* \text{and } w_{mid} = -(dA(x)/dx)\ U^\infty(z^*), \tag{8.64c}$$

which simply represent a vertical displacement of the streamlines through a distance—$\kappa_0$ A(x). The flow in the *upper deck* is driven by an outflow from the middle deck. Far from (8.64c) we have

$$\mathrm{Lim}_{z^* \to \infty} w_{mid}(x, z^*) = -dA(x)/dx. \tag{8.64d}$$

The flow expansions for the *upper deck* are

$$\begin{aligned} u &= 1 + \quad + \kappa_0{}^2 u_{up} + \dots \\ w &= \quad + \kappa_0{}^2 w_{up} + \dots \\ \pi &= \quad + \kappa_0{}^2 \pi_{up} + \dots \\ \theta &= \quad + \kappa_0{}^2\, \theta_{up} + \dots \end{aligned} \tag{8.65a}$$

Substituting in the equations of the local problem (8.60) shows that the motion is inviscid:

$$\begin{aligned}
&\partial u_{up}/\partial x + (1/\gamma)\partial \pi_{up}/\partial x = 0,\\
&\partial w_{up}/\partial x + (1/\gamma)\partial \pi_{up}/\partial z = (1/\gamma)B^*\theta,\\
&\partial u_{up}/\partial x + \partial w_{up}/\partial z = 0,\\
&\partial \theta_{up}/\partial x + B^*\left[(\gamma-1)/\gamma + (dT_\infty/dz_\infty)_{z_\infty=0}\right] w_{up},\\
&\left[\partial^2/\partial x^2 + \partial^2/\partial z^2 + K_0{}^2\right]\partial \pi_{up}/\partial x = 0,\\
&K_0{}^2 = (B^*/\gamma)\left[(\gamma-1)/\gamma + (dT_\infty/dz_\infty)_{z_\infty=0}\right].
\end{aligned} \tag{8.65b}$$

Note that in (8.64a) we have

$$\begin{aligned}
&\theta_{mid} = \chi(z^*),\ \text{and}\ \partial^2\pi_{mid}/\partial z^*\partial x = 0,\\
&\pi_{mid} = P(x) + Q(z^*),\\
&\theta_{mid} = (1/B^*)\, dQ/dz^* \equiv \chi(z^*),\\
&\chi(z^*) \equiv 0 \Rightarrow \theta_{mid} = 0, \pi_{mid} = P(x),\\
&\lim_{z\to 0}\pi_{up} \equiv \pi_{up}(x,\ 0) = P(x) \equiv \lim_{z^*\to\infty}\pi_{mid}.
\end{aligned} \tag{8.65c}$$

However, as a consequence of (8.64d) and matching with the upper deck,

$$w_{up}(x,\ 0) = -dA(x)/dx, \tag{8.65d}$$

and we can write the following relation between $\pi_{up}$ and A(x) :

$$\partial\left[\partial \pi_{up}/\partial z|_{z=0}\right]/\partial x = \gamma\left[K_0{}^2(dA/dx) + d^3A/dx^3\right]. \tag{8.65e}$$

This last relation (8.65e) is a boundary condition for the Helmoltz equation for $\pi_{up}$ in (8.65b). It is obvious that the middle deck solution (8.64c) does not satisfy the no-slip conditions on $\zeta = 0$, a situation which is remedied by the *analysis of the lower viscous deck below*, where the stretched variable is $\zeta = z/\kappa_0{}^2 \equiv z^*/\kappa_0$. Matching with the expansions (8.64a), when $z^* = 0(\kappa_0)$, implies the inner (lower) expansions:

$$\begin{aligned}
u &= k_0 u_{low} + \ldots\\
w &= \kappa_0{}^3 w_{low} + \ldots\\
\pi &= k_0{}^2 \pi_{low} + \ldots\\
\theta &= \theta_{low} + \ldots
\end{aligned} \tag{8.66a}$$

Substituting (8.66a) into the full equations of the local problem (8.60) yields the following *nonlinear viscous (BL) equations for* $u_{low}$, $w_{low}$, and $\theta_{low}$:

$$\begin{aligned}
&u_{low}\partial u_{low}/\partial x + w_{low}\partial u_{low}/\partial\zeta + (B^*/\gamma)\int_I (\partial\theta_{low}/\partial x)d\zeta\\
&\qquad +(1/\gamma)\, dP(x)/dx = \partial^2 u_{low}/\partial\zeta^2,\\
&\partial u_{low}/\partial x + \partial w_{low}/\partial\zeta = 0\,,\\
&u_{low}\partial\theta_{low}/\partial x + w_{low}\partial\theta_{low}/\partial\zeta = (1/Pr)\partial^2\theta_{low}/\partial\zeta^2
\end{aligned} \tag{8.66b}$$

The integration $\int_I$ is performed in the interval $[\infty, \zeta]$. For the system of equations (8.66b), we have the conditions:

$$\begin{aligned}&\zeta = 0 \ : \mathrm{u_{low}} = \mathrm{w_{low}} = 0, \theta_{\mathrm{low}} = \lambda_0 \Theta(\mathrm{x}),\ 0 \le \mathrm{x} \le 1,\\ &\zeta \to +\infty \ : \mathrm{u_{low}} \to \zeta, \mathrm{w_{low}} \to 0,\ \theta_{\mathrm{low}} \to 0, P(\mathrm{x}) \to 0,\\ &\mathrm{A(x)} \to 0, \mathrm{dA(x)/dx} \to 0 \text{ when } \zeta \to +\infty,\\ &\mathrm{x} \to -\infty \ : \mathrm{u_{low}} \to \zeta + \mathrm{A(x)},\\ &\mathrm{w_{low}} \to -\zeta \mathrm{dA(x)/dx},\ \theta_{\mathrm{low}} \to 0 \text{ when } \mathrm{x} \to -\infty,\end{aligned} \tag{8.66c}$$

after matching with the middle deck:

$$\mathrm{Lim}^x_{z\to o}\left[U^\infty(z^x) + k_o A(x)\frac{dU^\infty}{dz^x}\right] = \mathrm{Lim}_{\zeta\to\infty}(k_o u_{\mathrm{low}}); \tag{8.66d}$$

$$\mathrm{Lim}^x_{z\to o}\left[-k_o U^\infty(z^x)\frac{dA}{dz^x}\right] = \mathrm{Lim}_{\zeta\to\infty}\left(k_o^3 w_{\mathrm{low}}\right), \tag{8.66e}$$

and taking into account the fact that, $z^x = k_o\zeta$ :

$$U^\infty(z^x) \approx z^x, \text{ and } \frac{dU^\infty}{dz^x} \approx 1, \text{ when } z^x \to o. \tag{8.66f}$$

We note also that for $\pi_{\mathrm{low}}$ we have the expression:

$$\partial \pi_{\mathrm{low}}/\partial \mathrm{z} = B^* \theta_{\mathrm{low}} \ \Rightarrow \pi_{\mathrm{low}} = B^* \int_I \theta_{\mathrm{low}}\, \mathrm{d}\zeta + \mathrm{P(x)}. \tag{8.66g}$$

The *specification of the problem,* (8.66b)–(8.66g), *for the lower deck is completed by the relation* (8.65e), and the well-known interpretation of this relation is as follows: *the pressure driving the flow in the lower deck is itself induced in the main stream, i.e., in the upper deck, by the displacement thickness of the lower deck transmitted through the middle deck by the passive effect of displacement of the streamlines.*

The *strongly singular self-induced coupling* arises because the problem (8.66b) with (8.66c) to be solved in the lower viscous layer *does not accept* $P$(x) as data known prior to the resolution (as is the case in classical boundary layer problems). On the contrary, this pressure perturbation $P$(x) must be calculated at the same time as the velocity components $\mathrm{u_{low}}$ and $\mathrm{w_{low}}$, along with the temperature perturbation $\theta_{\mathrm{low}}$. Nevertheless, it should be emphasized that this function $P$(x) is not completely arbitrary, and that it is connected to the function A(x) through a relation.

This last relation is obtained via the analysis of perfect fluid flow in the upper layer [see (8.65e)]. If it is assumed that the parameter $\lambda_0$ in (8.66c) satisfies the condition

$$\lambda_0 << 1 \Rightarrow \tau_0 << \mathrm{M}_0, \tag{8.67}$$

then the equations (8.66b) may be linearized about the undisturbed boundary layer profile by making a further expansion.

Finally, we simply record here the solutions obtainable by using the Fourier transform in x, defined for the function f(x) by

$$< f(k) >= \int f(x)\exp(-ikx)\,dx, \text{ integrated between } (-\infty, \; +\infty).$$

In particular, we find instead of (8.65e), the relation

$$< A(k) >= i\left(N_0/\gamma\left[K_0{}^2 - k^2\right]\right) < P(k) > . \tag{8.68}$$

Where $N_0 = i\left[k^2 - K_0{}^2\right]^{1/2}$, if $|k| > K_0$ *and* $N_0 = i\left[K_0{}^2 - k^2\right]^{1/2}$, if $|k| < K_0$.

Here we have applied the standard *radiation condition* for choosing the sign of $N_0$, for $|k| < K_0$, when $z \to +\infty$, so that the *wave modes only carry energy upwards*. The definiton of $K_0{}^2$ is given in (8.65b).

## 8.8 Flow of a Thin Film Over a Rotating Disk

Thin liquid films flowing over a rotating disk surface provide a very interesting practical example for the application of our RAM approach. In the various experiments, the thin liquid film is mainly created by introducing deionized water from a flow collar at the center of an aluminum disk with a known initial film thickness and uniform radial velocity. The radial film thickness distribution is measured using a non-intrusive laser light interface reflection technique that enables the measurement of the instantaneous film thickness over a finite segment of the disk.

Experiments were performed for a range of flow rates, corresponding to Reynolds numbers (based on liquid inlet gap heights and velocities) between 238 and 1188. The angular speed of the disk was varied from zero revolutions per minute to 300 rev per minute. When the disk was stationary, a *circular hydraulic jump* was present in the liquid film, and the liquid-film thickness in the subcritical region (downstream of the hydraulic jump) was an order of magnitude greater than that in the supercritical region (upstream of the hydraulic jump), which was itself of the order of 0.3 mm. As the Reynolds number increased, the hydraulic jump migrated toward the edge of the disk.

In the case of rotation, the *liquid-film thickness* exhibited a *maximum* on the *disk surface*. The liquid-film inertia and friction influenced the *inner region*, where the *film thickness progressively increased.* The *outer region,* where the *film thickness decreased,* was primarily *affected by centrifugal forces.*

A *flow visualization study* of the thin film was also performed to determine the characteristics of the *waves on the free surface.* At *high* rotational speeds, *spiral waves* were observed on the liquid film. It was also determined that the angle of the waves which form on the liquid surface was a function of the ratio of the local radial to tangential velocity.

*Unsteady* liquid film flow on a rotating disk *can be analyzed by asymptotic methods for low and high Reynolds numbers, and* the analysis—in particular by *Brian G. Higgins*—elucidates *how a film of uniform thickness thins when the disk is set in steady rotation.* Below we refer mainly to the asymptotic results of *Higgins—but "à la Zeytounian", but we also make various complementary remarks.*

In the *low Reynolds number analysis, two time scales for the thinning film are identified.* The *long time analysis ignores the initial acceleration of the fluid layer* and hence is *singular at the onset of rotation.* The *singularity is removed by matching the long time scale expansion for the transient film thickness with a short time scale expansion that accounts for fluid acceleration during spinup.* The *leading order term in the long time scale solution for the transient film thickness is shown to be a lower bound for the film thickness for all time.*

A *short time analysis that accounts for boundary layer growth at the disk surface* is also presented for arbitrary Reynolds number. But, unfortunately, the analysis becomes invalid either when the boundary layer has a thickness comparable to that of the thinning film, or when nonlinear effects become important!

Gradual formation of a thin liquid film on a *hot/cold rotating disk* under the action of *thermocapillary force was also* studied analytically by *B.S. Dandapat and P.C. Ray, and* it was found that the *thermocapillary force acting on the free surface enhances the thinning rate for cooling the disk* (*see also Sect. 10.8* in our 2009 book [25], cited in "Introduction"). These authors consider the axisymmetric flow of a Newtonian fluid associated with the spreading of a thin liquid film on a rotating annular disk, with the effects of surface tension and gravity. The asymptotic solution for the free surface of the thin film is found by using an expansion for the film thickness in powers of a small parameter characterizing the thickness of the film and by applying the *method of matched asymptotic expansion.*

This solution can be used to calculate the thickness of the film, the velocity field, and the pressure at any point on the disk with good accuracy. Numerical results are presented for a specific initial distribution of the film thickness. Many features of the spin-coating thinning process are captured by their asymptotic solution.

### 8.8.1 A Mathematical Formulation

The unsteady flow of a thin liquid film on a *rotating disk is analysed by means of matched asymptotic expansion under the assumption of a radially uniform film thickness h that varies with time h = h(t),* and the initial film thickness $h_0$.

Below, as is usual in many papers, the *starting equations take into account the von Karman similarity form (1921)*. In the *thinning film, we restrict our attention to flows that maintain a film thickness that is independent of the radial coordinate r and azimuthal coordinate* θ as *dimensional (Navier) starting* equations, we write, for an *incompressible Newtonian fluid*:

$$\begin{aligned}
&\partial u/\partial r + u/r + \partial u/\partial z = 0,\\
&\partial u/\partial t + u\partial u/\partial r - v^2/r + w\partial u/\partial z + (1/\rho_0)\partial p/\partial r\\
&\quad = \nu_0\left\{\partial^2 u/\partial r^2 + \partial(u/r)/\partial r + \partial^2 u/\partial z^2\right\},\\
&\partial v/\partial t + u\partial v/\partial r + u\,(v/r) + w\partial v/\partial z\\
&\quad = \nu_0\left\{\partial^2 v/\partial r^2 + \partial(v/r)/\partial r + \partial^2 v/\partial z^2\right\},\\
&\partial w/\partial t + w\partial w/\partial z + u\partial w/\partial r + (1/\rho_0)\partial p/\partial z\\
&\quad = \nu_0\left\{\partial^2 w/\partial r^2 + (1/r)\partial w/\partial r + \partial^2 w/\partial z^2\right\}.
\end{aligned} \tag{8.69a}$$

At *the surface z = 0 of the rotating disk*, the adherence and impenetrability conditions stipulate that

$$u(r,0,t) = 0,\ v(r,0,t) = r\,\Omega_0,\ w(r,0,t) = 0, \tag{8.69b}$$

while at *the free surface z = h(t)*, the normal and tangential components of the traction vector must vanish, and in addition, we have the kinematic condition:

$$\begin{aligned}
&-p + 2\mu_0\partial w/\partial z = 0,\\
&\partial w/\partial r + \partial u/\partial z = 0,\ \partial v/\partial z = 0,\\
&dh/dt = w(r,h,t).
\end{aligned} \tag{8.69c}$$

The initial conditions for the above equations (8.69a) are

$$u = v = w = 0,\ h = h_0,\ dh/dt = 0 \text{ at } t = 0. \tag{8.69d}$$

*Equations* (8.69a), *with initial* (8.69d) *and boundary conditions* (8.69b) *and* (8.69c), *define a typical initial-boundary value problem!*

### 8.8.2 The Reduced Initial Boundary-Value Problem à la von Karman

We will seek a solution to the above I-BCV problem in the following form:

$$u = rf(z,t),\ v = rg(z,t),\ w = w(z,t). \tag{8.70a}$$

Substituting this in, (8.69a) we obtain the reduced system of equations

$$\begin{aligned} &2f + \partial w/\partial z = 0\,. \\ &\partial f/\partial t + f^2 - g^2 + w\partial f/\partial z = \nu_0 \partial^2 f/\partial z^2, \\ &\partial g/\partial t + 2fg + w\partial g/\partial z = \nu_0 \partial^2 g/\partial z^2, \end{aligned} \tag{8.70b}$$

and for the reduced equations (8.70b), we can write as conditions:

$$\begin{aligned} &f(0,t) = 0, g(0,t) = \Omega_0,\ w(0,t) = 0, \\ &\partial f/\partial z = 0,\ \partial g/\partial z = 0,\ \text{at } z = h,\ \text{and } dh/dt = w(h,t), \\ &f(z,0) = 0, g(z,0) = 0, w(z,0) = 0,\ h = h_0,\ dh/dt = 0,\ \text{ at } t = 0. \end{aligned} \tag{8.70c}$$

We observe that the kinematic condition $dh/dt = w(h,t)$ can be used to determine the *film thickness h(t)*. The pressure $p$ can be determined from

$$-p + 2\mu_0 \partial w/\partial z = 0 \ \text{ at the free surface } z = h(t), \tag{8.70d}$$

*once* w *and h are known!*

### 8.8.3 Dimensionless RAM Approach

In the RAM approach, one must first carry out the dimensionless analysis, introducing the following dimensionless variables:

$$\tau = \tau/t_c, \zeta = z/h_0,\ H = h/h_0, \tag{8.71a}$$

$$F = f/(1/t_c),\ W = w/(h_0/t_c),\ G = g/\Omega_0, \tag{8.71b}$$

where the choice of reference time $t_c = \nu_0/(h_0\,\Omega_0)^2$ is obtained by requiring that, in the limit of a *vanishingly small Reynolds* number ($Re = h_0^2\Omega_0/\nu_0$), the *centrifugal force be balanced by viscous shear across the film.* This time $t_c$ may be interpreted as the time required after *spinup* for the film to thin, in the absence of inertial forces other than the *centrifugal* force!

In fact, we observe that the time scale for vorticity generated at the disk boundary to spread through the film on spinup is $h_0^2/\nu_0$, whereas the time scale for inertial spinup is $1/\Omega_0$.

In terms of above (8.71a) and (8.71b), we obtain the following dimensionless problem for F, G, W, and H:

$$
\begin{aligned}
&2F + \partial W/\partial\zeta = 0\,,\\
&Re^2\{\partial F/\partial\tau + F^2 + W\partial F/\partial\zeta\} - G^2 = \partial^2 F/\partial\zeta^2,\\
&Re^2\{\partial G/\partial t + 2FG + W\partial G/\partial\zeta\} = \partial^2 G/\partial\zeta^2,\\
&F(0,\tau) = 0, G(0,\tau) = 1,\ W(0,\tau) = 0,\\
&\partial F/\partial\zeta = 0,\ \partial G/\partial\zeta = 0,\ \text{at } \zeta = H,\\
&dH/d\tau = W(H,\tau),\\
&F(\zeta,0) = 0, G(\zeta,0) = 0,\ W(\zeta,0) = 0,\ H = 1,\ dH/d\tau = 0,\ \text{at } \tau = 0.
\end{aligned}
\tag{8.72}
$$

We observe that the above initial-boundary value problem (8.72) is only really well suited for the outer *long time scale limit analysis* described below.

### 8.8.4 Outer Long Time Scale Limit: Re→0 with τ and ζ Fixed Zero-Order Outer Problem

At the leading zero-order, when

$$\text{Re} \to 0 \text{ with } \tau \text{ and } \zeta \text{ fixed,} \tag{8.73a}$$

the above problem (8.72) is very degenerate at $O(Re^0)$:

$$
\begin{aligned}
&2F_L{}^0 + \partial W_L{}^0/\partial\zeta = 0\,,\\
&\partial^2 F_L{}^0/\partial\zeta^2 + G_L{}^{02} = 0\,,\\
&\partial^2 G_L{}^0/\partial\zeta^2 = 0\,,\\
&F_L{}^0(0,\tau) = 0, G_L{}^0(0,\tau) = 1, W_L{}^0(0,\tau) = 0\,,\\
&\partial F_L{}^0/\partial\zeta = 0,\ \partial G_L{}^0/\partial\zeta = 0,\ \text{at } \zeta = H_L{}^0.
\end{aligned}
\tag{8.73b}
$$

As a consequence, the *initial conditions* at $\tau = 0$ where the functions carry a lower subscript <L>, *must be discarded* at this level of approximation, and an inner short-time-scale solution will be required to satisfy these initial conditions.

The solution of the reduced outer problem (8.73b) for $F_L{}^0$, $G_L{}^0$ and $W_L{}^0$ is

$$
\begin{aligned}
&F_L{}^0(\zeta, H_L{}^0) = -\zeta^2/2 + H_L{}^0\zeta,\\
&G_L{}^0(\zeta, H_L{}^0) = 1,\\
&W_L{}^0(\zeta, H_L{}^0) = \zeta^3/3 - H_L{}^0\zeta^2.
\end{aligned}
\tag{8.73c}
$$

### 8.8.5 First-Order Outer Problem

Because the right-hand side in the corresponding equations is known from the above zero-order solution (8.73c), the first-order problem emerging from (8.72) at $O(Re^2)$ can be expressed, and the *kinematic condition* $dH/d\tau = W(H,\tau)$ in (8.72) provides the additional relation needed to determine the film thickness as a function of time $\tau$. Hence, we have

$$H0(\tau) = [3/(4\tau + C_0)]^{1/2}, \quad H1(\tau) = C_1(H0(\tau))^3 - (17/105)(H0(\tau))^5. \tag{8.73d}$$

The constants of integration C0 and C1 are to be determined from the initial conditions [see the last line in (8.72)].

### 8.8.6 Inner Short Time Scale Analysis

In the present case, instead of $\tau$, an obvious new short time scale is

$$T = \tau/Re^2, \tag{8.74}$$

and the appropriate dimensionless functions for the short time scale analysis are such that, instead of the dimensionless problem (8.72), we consider the following *local problem near the initial time* $T = 0$, for $F_{sh}$, $G_{sh}$, $W_{sh}$, and $H_{sh}$:

$$\begin{aligned}
&2F_{sh} + \partial W_{sh}/\partial\zeta = 0\,,\\
&\partial F_{sh}/\partial T + Re^2\left[{F_{sh}}^2 + W_{sh}\partial F_{sh}/\partial\zeta\right] - {G_{sh}}^2 = \partial^2 F_{sh}/\partial\zeta^2,\\
&\partial G_{sh}/\partial T + Re^2[2F_{sh}G_{sh} + W_{sh}\partial G_{sh}/\partial z] = \partial^2 G_{sh}/\partial\zeta^2\\
&F_{sh}(0,T) = 0, G_{sh}(0,T) = 1,\ W_{sh}(0,T) = 0,\\
&\partial F_{sh}/\partial\zeta = 0,\ \partial G_{sh}/\partial\zeta = 0,\ \text{at } \zeta = H_{sh},\\
&dH_{sh}/dT = W(H_{sh},T),\ H_{sh} = 1,\ dH_{sh}/dT = 0,\ \text{at } T = 0,\\
&F_{sh}(\zeta,0) = 0, G_{sh}(\zeta,0) = 0,\ W_{sh}(\zeta,0) = 0,\ \text{at } T = 0.
\end{aligned} \tag{8.75}$$

### 8.8.7 Zero-Order Local Short Time Scale Problem

The *leading order,* when

$$Re \to 0,\ \text{with T and } \zeta \text{ fixed}, \tag{8.76a}$$

follows by *setting the terms multiplied by* $Re^2$ *in the second and third equations of* (8.75) *equal to zero*. This yields a linear set of unsteady partial differential equations, and an important consequence of the local-in-time limiting process is that the film thickness at zero time can be determined directly from the kinematic condition:

$$H_{sh}{}^{0} = 1, \quad dH_{sh}{}^{0}/dT = 0, \quad \text{at } T = 0 \tag{8.76b}$$

Hence, in the zero-order approximation, the axial coordinate z is specified only in the range $0 \leq \zeta \leq 1$. For $G_{sh}{}^{0}(\zeta,T)$, we derive the following zero-order local short time scale problem, with initial condition $G_{sh}{}^{0}(\zeta,0) = 0$:

$$\begin{aligned} &\partial G_{sh}{}^{0}/\partial T = \partial^2 G_{sh}{}^{0}/\partial \zeta^2,, 0 < \zeta < 1,\ T > 0,. \\ &G_{sh}{}^{0}(0,T) = 1, \\ &G_{sh}{}^{0}(1,T) = 0, \end{aligned} \tag{8.76c}$$

while for $F_{sh}{}^{0}(\zeta,T)$, with the initial condition $F_{sh}{}^{0}(\zeta,0) = 0$, we have the following problem:

$$\begin{aligned} &\partial F_{sh}{}^{0}/\partial T - \partial^2 F_{sh}{}^{0}/\partial \zeta^2 = G_{sh}{}^{02}, 0 \leq \zeta \leq 1,\ T > 0, \\ &F_{sh}{}^{0}(0,T) = 1, \\ &F_{sh}{}^{0}(1,T) = 0. \end{aligned} \tag{8.76d}$$

For a solution of the above zero-order local short time scale problem, see pp. *3526–3527 of* the *paper by Brian G. Higgins in Physics of fluids 29(11).*

However, to determine the constants $C_0$ and $C_1$ in (8.73d), one must compute the film thickness for the short time scale solution to (first) order $Re^2$. To do this, we introduce an *intermediate time scale*

$$\sigma = Re^{2\alpha} T, \text{ with } 0 < \alpha < 1, \tag{8.76e}$$

and then *match up* the expressions for $H_L$ and $H_{sh}$ in their region of overlap. In this way, a *single composite uniform asymptotic expansion for the transient film thickness is derived* by <*adding* the *long time scale solution* to the *short time scale solution and subtracting their common part from the result*>:

$$H^c(\tau; Re) = H_L{}^{0}(\tau) + (2/3)\tau + Re^2\left[H_{sh}{}^{1}(\tau/Re^2) + H_L{}^{1}(\tau)\right] - 0.6609. \tag{8.77}$$

### *8.8.8 Complementary Remarks*

The underlying assumption in developing the short-time solution is that the effects of viscosity are felt throughout the film on spinup. At low $Re \ll 1$, this seems a reasonable assumption, but at large Re $\gg 1$, we would anticipate a boundary layer to form at the disk surface, in which the effects of viscosity are felt. The boundary layer thus provides a pathway for liquid to leave the disk during spinup, with liquid being supplied to it from an overlying inviscid layer.

In Sect. IV of the Higgins paper—*inspired by C.J. Lawrence et al.*—the inner solution is analyzed, assuming that the characteristic time scale $t_c$ for the flow is $\varepsilon/\Omega_0$, where $\varepsilon \ll 1$. But it appears that *matching* with an outer solution *is best accomplished numerically*. This conclusion seems reasonable in the framework of our RAM approach! Simpler reduced working models need to be derived, but these models must be relevant, and they must belong to a hierarchy of working models derived rationally from a well-posed *initial-boundary value problem*.

*In fact, the accurate derivation of working models is the first, and very important key step. Only then is it worth trying to solve these derived working models numerically, to gain further insight into film dynamics at low or high Re !*

The composite expansion *à la Higgins* predicts a film of zero thickness when the spinning time becomes indefinitely large, a result that is obviously *physically unreasonable*! *Not accounted* for in this analysis are *interface effects* that eventually *dominate* the *hydrodynamics* at the rim of the disk when the *film becomes sufficiently thin* (*Marangoni effect*). The effect of gravitational forces tends to promote film thinning, while the other two effects, inertial and surface tension forces, tend to retard it! On the other hand, although a flat interface is the preferred shape during spinning at low Re, the approximation may well be inappropriate at high Re! Circumferential waves and helical waves are known to occur, although a detailed analysis of their dynamics has never been undertaken.

Concerning the *unsteady flow of a thin liquid film on a cold/hot rotating disk, when the thermocapillary force induced by the variation in the surface tension with temperature is taken into account, the reader can find some information in Sect. 10.8 of* our 2009 book [25], cited in "Introduction".

## 8.9 The KZK–Parabolic Single Model Equation in Nonlinear Acoustics

*Acoustics is concerned with the study of small oscillations in fluids (liquids* and gases). The traditional study of acoustics uses with the linearized equations of fluid mechanics. However, below, we discuss fluctuations that violate the assumptions of linearity.

The fundamental equations of nonlinear acoustics are those of fluid dynamics, and the mathematical description begins with continuity equations.

*Nonlinear acoustics* is a branch of physics and acoustics dealing with sound waves of sufficiently large amplitude, and this means one must use the complete system of fluid dynamics equations (for sound waves in liquids and gases). These equations are generally nonlinear, and their traditional linearization is no longer possible. The solutions of these equations show that, due to the effects of nonlinearity, sound waves are distorted as they travel.

A sound wave propagates through a material as a localized pressure change. Increasing the pressure of a gas, or fluid, increases its local temperature, and the local speed of sound in a compressible material increases with temperature. As a result, the wave travels faster during the high pressure phase of the oscillation than during the lower pressure phase. The general wave equation that accounts for nonlinearity up to the second order is the *Westervelt equation.*

The *Westervelt equation* can be simplified to a one-dimensional form. With an assumption of strictly forward propagating waves and the use of a coordinate transformation to a retarded time frame, this corresponds to the *famous viscous Burgers equation.* An *extension* of the Burgers equation that accounts for the *combined effects of nonlinearity, diffraction, and absorption in directional sound beams* is the *Khokhlov-Zabolotskaya-Kuznetsov* (KZK) equation. Solutions to this equation are generally *used to model nonlinear acoustics.*

## 8.9.1 The RAM Approach

Here we give only a *very short, but rational derivation, along the lines of the RAM approach,* starting from the dominant dimensionless acoustic system of NSF equations given in Sect. 2.4 by the equations (2.9a–2.9d) for the *velocity vector* **u** and the *thermodynamic perturbations* $\pi$, $\omega$, and $\theta$, relative to unity, *neglecting* the terms proportional to $O(M^2)$. In these equations (2.9a and 2.9d), $Re_{ac}$ is the *acoustic Reynolds number, which* is always *very large* compared to unity. We may thus say that *the medium is weakly dissipative, at the chosen frequency.* In fact, we have a *perturbation problem* with *two small parameters,* M and $1/Re_{ac}$, but it is also *necessary* to take into account the fact that *the parabolic approximation, which leads to the KZK model equation, may not be valid close to the source,* in the *near acoustic field!*

The condition for this approximation to be valid is that *the width of the acoustic source satisfy*

$$d >> (1/k_{ac}), \tag{8.78a}$$

where the *wave number* $k_{ac}$ is associated with the reference frequency $\omega_{ac}$ according to $\omega_{ac}/k_{ac} =$ *sound velocity.*

As a consequence,

$$\alpha = (1/k_{ac})/d << 1 \tag{8.78b}$$

is the *third small parameter* for the derivation of the KZK model equation, and one must assume the existence of *two similarity relations*:

$$1/\mathrm{Re}_{ac} = \kappa\,M, \quad \alpha^2 = \beta\,M, \quad \text{with } \kappa = O(1) \text{ and } \beta = O(1). \tag{8.78c}$$

The KZK equation is only significant (asymptotically) in the *far acoustic field*, at *large distances from the acoustic source*, and for this, we introduce *two new dimensionless coordinates* $\eta$ and $\zeta$ defined by

$$\eta = \alpha y \text{ and } \zeta = \alpha z, \text{ with } \alpha \downarrow 0 \text{ as } M \downarrow 0. \tag{8.78d}$$

We observe that, since the width of the acoustic source d should be much greater than the wavelength 1/k, the transverse field variations are slow compared with longitudinal variations along the acoustic axis. With (8.78b) we also write

$$V = v/\alpha, \; W = w/\alpha, \text{ such that } \mathbf{u} = u\,\mathbf{i} + \alpha(V\,\mathbf{j} + W\,\mathbf{k}), \tag{8.78e}$$

However, one must assume that the functions u, V, W, $\pi$, $\omega$, $\theta$ depend on t, x, $\eta$, $\zeta$, and also the low Mach number $M \ll 1$, at least when we take into account the above two similarity rules (8.78c).

Now, from the transformed equations for u, V, W, $\pi$, $\omega$, $\theta$, as dependent functions of t, x, $\eta$, $\zeta$ and M, instead of the system of equations (2.9a–2.9d), we have the low Mach number asymptotic expansion (8.78f), with t, x, $\eta$, $\zeta$, and Pr, $\kappa$, $\beta$ fixed:

$$U \equiv (u, V, W, \pi, \omega, \theta) = U_0 + M\,U_1 + \ldots. \tag{8.78f}$$

### 8.9.2 *Leading Order System for $U_0$ in the Simplistic Tentative Expansion (8.78f)*

At *leading order*, the *acoustic system* for $u_0$, $\omega_0$, $\pi_0$, and $\theta_0$ is

$$\begin{aligned} &\partial\omega_0/\partial t + \partial u_0/\partial x = 0,\\ &\partial u_0/\partial t + (1/\gamma)\partial\pi_0/\partial x = 0,\\ &\partial\theta_0/\partial t + (\gamma - 1)\partial u_0/\partial x = 0,\\ &\pi_0 = \theta_0 + \omega_0. \end{aligned} \tag{8.78g}$$

For $V_0$ and $W_0$, we have *two equations*, as a *consistent consequence of our RAM approach*:

$$\partial V_0/\partial t + (1/\gamma)\partial\pi_0/\partial\eta = 0, \quad \partial W_0/\partial t + (1/\gamma)\partial\pi_0/\partial\zeta = 0, \tag{8.78h}$$

or

$$\partial/\partial t[\partial V_0/\partial \zeta - \partial W_0/\partial \eta] = 0. \tag{8.78i}$$

It is interesting to observe that the last equation (8.78i) replaces the *physically unrealistic assumption on p. 335* of *Coulouvra*t <2>, cited in Chap. 2, where an "ad hoc" *approach is used* (for the *dissipative case*), applying the *potential equation* $\mathbf{u} = -\nabla\Phi$, *with the assumption that the motion is irrational*, which *is obviously superfluous*!

From the leading-order acoustic system (8.78g) for, $u_0$, $\omega_0$, $\pi_0$, and $\theta_0$, we derive for $u_0$ the classical equation

$$\partial^2 u_0/\partial t^2 - \partial^2 u_0/\partial x^2 = 0, \text{ and } u_0 = F(\chi, \eta, \zeta), \text{ with } \chi = t - x. \tag{8.79a}$$

However, such a solution for $u_0$ is *a good solution only for the near acoustic field, close to the acoustic source, while the KZK equation is on the contrary a far-field equation, holding far from the acoustic source!*

To avoid various *cumulative effects,* one must consider (8.78f) as a *two-scale expansion* and assume

$$u_0 = F^*(\chi, \xi, \eta, \zeta), \text{ with } \xi = Mx. \tag{8.79b}$$

We also observe that $\partial/\partial t = \partial/\partial\chi$, whence we obtain

$$\begin{aligned}\pi_0 &= \gamma F^*(\chi, \xi, \eta, \zeta),\\ \omega_0 &= F^*(\chi, \xi, \eta, \zeta),\\ \theta_0 &= (\gamma - 1)\, F^*(\chi, \xi, \eta, \zeta),\end{aligned} \tag{8.79c}$$

and in place of (8.78h),

$$\partial V_0/\partial\chi + \partial F^*/\partial\eta = 0 \text{ and } \partial W_0/\partial\chi + \partial F^*/\partial\zeta = 0. \tag{8.79d}$$

### 8.9.3 Second-Order System for $U_1$

From the *second-order system of equations* for $u_1$, $\omega_1$, $\theta_1$, $\pi_1$ and (8.79d), we obtain a single inhomogeneous acoustic equation for $u_1$ with a source term, viz.,

$$\partial^2 u_1/\partial t^2 - \partial^2 u_1/\partial x^2 = \partial\Lambda(F^*)/\partial\chi + (\beta/\gamma)\left[\partial^2 F^*/\partial\eta^2 + \partial^2 F^*/\partial\zeta^2\right], \tag{8.80a}$$

where, for the function $\Lambda(F^*)$, we have the following rather unexpected and complicated expression:

$$\Lambda(F^*) = \kappa[4/3 + (\gamma - 1)/\mathrm{Pr}]\partial^2 F^*/\partial\chi^2 + (\gamma + 1)F^*(\partial F^*/\partial\chi) - 2\partial F^*/\partial\xi. \tag{8.80b}$$

### 8.9.4 The Compatibility-Non-Secularity Condition

Finally, we consider the *compatibility-non-secularity condition, a classical requirement in the multiple scale approach which provides a way to cancel terms that do not remain bounded in the solution for* $u_1$, *because these* terms *will ultimately be greater than* $u_0$, *however small the Mach number!* The only way for the above low Mach number asymptotic expansion (8.78f) to be *non-secular* is for the following *compatibility relation to be valid, according to* (8.80a):

$$\partial \Lambda(F^*)/\partial \chi + (\beta/\gamma)\left[\partial^2 F^*/\partial \eta^2 + \partial^2 F^*/\partial \zeta^2\right] = 0. \tag{8.80c}$$

### 8.9.5 KZK Equation

Using (8.80c) in (8.80b), we obtain the following KZK Model equation:

$$\begin{aligned}&\partial^2 F^*/\partial \chi \partial \xi - (\beta/2\gamma)\left[\partial^2 F^*/\partial \eta^2 + \partial^2 F^*/\partial \zeta^2\right]\\ &- [(\gamma+1)/2]\partial/\partial \chi[F^* (\partial F^*/\partial \chi)]\\ &= S^* \partial^3 F^*/\partial \chi^3,\end{aligned} \tag{8.81a}$$

with

$$S^* = (\kappa/2)[4/3 + (\gamma - 1)/Pr]. \tag{8.81b}$$

This elegant KZK equation (8.81a) simultaneously and consistently takes into account the *nonlinearity*, *dissipation*, and *diffraction*.

The KZK equation is by far the most useful equation, owing to its (*relative*) *simplicity*, and *practically all numerical schemes* dealing with *strongly nonlinear propagation* rely on it, even for rather wide-band signals and interaction within the near field of piston sources.

The KZK equation (8.81a) with (8.81b) can also be written in the *Coulouvrat form*

$$\begin{aligned}&\partial^2 F^*/\partial \chi \partial \sigma - [N/4]\nabla_\perp^2 F^x \equiv \frac{\partial^2 F^x}{\partial \eta^2} + \frac{\partial^2 F^x}{\partial \zeta^2} - [1/\Gamma]\partial^3 F^*/\partial \chi^3\\ &= \partial/\partial \chi[F^* (\partial F^*/\partial \chi)],\end{aligned} \tag{8.82}$$

with

$$N = 4\beta/\gamma(\gamma - 1) \text{ and } \Gamma = 2S^*/(\gamma + 1). \tag{8.83}$$

and

$$\nabla_{\perp}^{2} F^{x} \equiv \frac{\partial^{2} F^{x}}{\partial \eta^{2}} + \frac{\partial^{2} F^{x}}{\partial \zeta^{2}} \tag{8.84}$$

For the far field,

(I) when $\Gamma << 1$ and $\mathrm{N}\Gamma >> 1$, we have thermoviscous attenuation, and
(II) when $\Gamma >> 1$ and $\mathrm{N} << 1$, we have nonlinear attenuation as the leading effect.

The case (I), when the Goldberg number $\Gamma$ is small ( $<< 1$), plays an important role for parametric acoustic source emission.

Finally, we observe that experimental values show that $\Gamma$ and N remain roughly of order 1, which proves that the hypotheses made in the asymptotic derivation of the above KZK equation (8.81a) with (8.81b) are consistent!

## References

<1> Godts S, Zeytounian RKh (1991) J Eng Math 28:93–98
<2> Guiraud JP, Zeytounian RKh (1976) A theory for the core of high vortex sheets. ONERA TP N° 89
<3> Guiraud JP, Zeytounian RKh (1982) Vortex sheets and concentrated vorticity: a variation on the theme of asymptotic modelling in fluid mechanics. In: Hormung Hg, Müller EA (eds) Proceedings of a Coloquium "Vortex Motion" held in Goettingen on the occasion of the 75th anniversary of the Aerodynamishe Versuchsanstalt in November 1982. Vieweg and Son, Braunschweig, pp 31–49
<4> Zeytounian RKh (1993) Nonlinear long surface waves in shallow water. Unpublished Typescript. University of Lille-I, LML, 224p
<5> Zeytounian RKh (1991) A quasi-one-dimensional asymptotic theory for nonlinear water waves. J Eng Math 28:262–296
<6> Zeytounian RKh (1995) Nonlinear long waves on water and solitons. Physics-Uspekhi 38(12):1333–1381
<7> Godts S, Zeytounian RKh (1990) ZAMM 70(1):67–69

# Chapter 9
# Some Concluding Remarks about Part III

## 9.1 RAM Approach

*First, concerning our RAM approach, I would like to stress once again* that *the three capital letters are important: rationality* as *opposed* to an *ad-hoc* approach, *asymptotics* to *emphasize* the fact that *perturbative methods* stand at the core of the theory, and taking these *together* as the *two main working concepts,* we obtain a m*odeling* approach to fluid dynamics, *the main goal being specifically to provide useful reduced models accessible to numerical simulation via high-speed computing.*

This approach has contributed to a *fresh understanding* of *Newtonian fluid flow problems* and has *opened up new avenues for tackling real stiff fluid flow phenomena*, which are known to lead to very difficult problems, not only mathematically, but also often numerically in computing.

*Turbulent fluid flow problems definitely lie outside the scope of the present book, as* was already *clearly* mentioned in the *preface.*

I think the reader interested in the RAM approach should be able to apply this approach to specific problems in fluid dynamics using the framework provided by the *nine examples* considered in Chap. 8. The curious reader will find substantial further assistance in our books, published by Springer in Heidelberg during the years *1986–1987*, 1990, 1991, 1994, 2002, 2004, 2006, 2009, and 2012. Here a wide range of simplified but accurately reduced working models are given, emerging from the *Euler* (non-viscous), Navier (incompressible), and *NSF* (dissipative) equations and *ready* for *numerical simulation/computation.*

*This is obviously not an easy task, but it is a necessary one if we wish to come up with a rational and relevant modeling process for consistent numerical simulation on a high speed computer! We mention some of these examples below.*



R.Kh. Zeytounian, *Challenges in Fluid Dynamics*,
DOI 10.1007/978-3-319-31619-2_9

## 9.2 Hydrodynamic Stability Theory

In the *book edited in 2004* [14], cited in "Introduction", with the title *Theory and Applications of Viscous Fluid Flows, Chap. 9 discusses some aspects of* hydrodynamic stability theory, and in particular, the *Guiraud-Zeytounian asymptotic approach to nonlinear hydrodynamic stability, which involves a decomposition in normal modes and expansion techniques. This was an idea introduced* by Stuart and Watson. Unfortunately, when the modes are not discrete, but spread over a continuum, the *Stuart-Watson technique fails*!

In 1978, Guiraud and I succeeded in filling this gap, in a purely formal way, for continuously spread *Tollmien-Schlichting waves,* with a technique similar to the one used for discrete modes. In fact, our approach provides an economical way to derive the equation or set of equations *that govern the evolution of the amplitude of the most rapidly amplified modes in the linear theory.*

We observe that the *small parameter*, which may used as a starting point for *running the machinery of the asymptotic modeling*, is simply *the width of that area compared with the reciprocal of the length which characterizes the width of the transverse distribution of velocity being tested against stability.* Asymptotic modeling allows one to follow the fate of a perturbation from the initial stages, when it is widely spread across all the eigenmodes, to the final stages of its nonlinear evolution.

In the case of confined perturbations, in the framework of a nonlinear theory, we derive an amplitude equation which can be written in the form

$$\mathrm{dA_n}/\mathrm{d}\tau - L\mathrm{A_n} \; - \Sigma_{\mathrm{Rn}} P\mathrm{M}\big(\mathrm{A}_p,\, \mathrm{B}_{q,r}\big) = 0, \tag{9.1}$$

where $P$ is a projection operator and M(a, b) is a *bilinear* function of two variables. In (9.1), $\Sigma_{\mathrm{Rn}}$ is the sum over all resonant (p, q, r) triplets, where $\mathrm{R_n}$ is a resonance condition ($p + q + r = \mathrm{n}$).

In the particular case of only one element, we have the *exchange of stabilities*, and $\mathrm{A_n}$ is well defined by one real scalar $C(\tau)$ which must be a solution of the classical *Landau-Stuart equation*:

$$\mathrm{d}C/\mathrm{d}\tau = \lambda^{\circ} C + \mathrm{K}C^3, \tag{9.2}$$

Concerning the stability of *Rayleigh-Bénard thermal convection* between two infinite parallel planes as boundaries, we show that nonlinearity occurs through quadratic interactions between modes that have wave-number vectors of equal lengths, but rotated through $\pi/3$. In particular, a careful examination shows that the *statement by Schüler et al. (1965)* concerning the *vanishing of the quadratic interaction in the nonlinear interaction of a finite number of discrete modes is true only for boundaries that are either both rigid or both free!*

The *nonlinear* hydrodynamic stability *theory* is investigated for *confined* and *unconfined* perturbations *in Chap. 9* of [14], cited in "Introduction", and we merely stress that our treatment of unconfined flow with Guiraud is quite parallel to the treatment of confined flow. When one recalls that this last treatment may be justified by a rigorous mathematical argument, it may be enlightening to know that, formally at least, a quite analogous treatment also exists for unconfined flow.

## 9.3 Bénard Thermal Convection

The reader will find some complementary information concerning the modeling of the *Bénard heated from below thermal convection* problem in our 2009 book [25], cited in "Introduction". In particular, to a certain extent, the short *Chap. 8,* with a *summary* of Chaps. 3, 5, and 7 of [25], cited in "Introduction", responds to a suggestion by Professor René Moreau (Series Editor of FMIA—in 2009):

H*owever, from a certain point of view, you know that some readers do not care much for this rigor and just want to know what the relevant model equations are for their problem!*

On pp. 359–363 of [25], cited in "Introduction", the reader will also find the RAM derivation of the so-called *anelastic (deep) equations.* These equations are derived again in Sect. 9.4, with some modifications and comments!

## 9.4 Anelastic (Deep) Equations

The full NSF equations are rarely considered in recent papers devoted to anelastic systems of equations, and in fact, *the various derivations are systematically performed in an ad hoc manner*! The *anelastic equations* considered below are *non-adiabatic* and *viscous heat conducting equations for atmospheric thermal convection,* treated *à la Zeytounian. T*he Boussinesq number B0 is assumed to be $\approx 1$, and this *deep system of equations* is a working model, emerging from the full NSF system of equations, that is valid *for the whole troposphere.*

The derivation of these *anelastic deep dissipative* equations using the RAM approach is very instructive, because one must carry out a careful inspection of the effects of the various terms in the full NSF atmospheric starting equations, so this is a very interesting test for our RAM approach. Below we describe some of the steps in this derivation!

The starting *hypothesis* concerns the Boussinesq number:

$$\mathrm{Bo} = \mathrm{H}_0/(\mathrm{RT}_{\mathrm{st}}(0)/g) \approx 1, \tag{9.3a}$$

where $\mathrm{T}_{\mathrm{st}}(0)$ is the temperature of a standard atmosphere on the ground $z = 0$, and $\mathrm{H}_0$ is the vertical scale of the considered atmospheric motion in the troposphere, with

$$H_{st} \approx RT_{st}(0)/g \approx 10^4 m, \tag{9.3b}$$

The full unsteady NSF atmospheric equations are written in dimensionless form for the velocity vector **u**, and thermodynamic perturbations $\pi$, $\omega$, $\theta$ *relative to the standard atmosphere*, assumed to be a function of $z_{st} = \mathrm{Bo}\, z$, where $z_{st}$ is the altitude of the standard atmosphere.

For the perturbation $\theta$ of the temperature T, given by

$$\theta = [T - T_{st}(z_{st})]/T_{st}(z_{st}), \text{ on the ground } z = 0, \tag{9.3c}$$

we write the *thermal dimensionless condition*

$$\theta = \tau_0 \Theta(t, x, y) \text{ on } z = 0, \tag{9.4}$$

with $\tau_0 = \Delta T_0/T_{st}(0)$, and where $\Delta T_0$ is a thermal measure associated with the given (dimensionless) distribution function $\Theta(t, x, y)$ on the ground $z = 0$ in a limited domain. In fact, this parameter $\tau_0$ in (9.4) is related to the above Boussinesq number Bo, and also to the parameter $h_0 = R\Delta T_0/g$:

$$\tau_0 = \mathrm{Bo}/\nu_0, \text{ with } \nu_0 = H_{st}/h_0. \tag{9.5}$$

The parameter $h_0 << H_{st}$ is related to the vertical dimensionless *displacement of a particle* $\delta$ at a fixed dimensionless time t, with respect to its position in the standard atmosphere. In fact, with the parameter $\nu_0 >> 1$, one must also consider the *Froude* number

$$\mathrm{Fr}_{L0} = U°/(L_0 g)^{1/2}, \tag{9.6}$$

where $L_0$ is a horizontal scale for the atmospheric motion under consideration.

Below we consider a *local mesoscale motion*, such that

$$\varepsilon = H_0/L_0 \approx 1 \Rightarrow \mathrm{Ro} >> 1. \tag{9.7}$$

The *Rossby number* $\mathrm{Ro} = U_0/L_0 f_0$, as defined by (2.13c), is $\approx 1$ only if $L_0 \approx 10^5$m, since $f_0 \approx 10^{-4}$, for $U_0 \approx 10$m/s, with (9.7).

Next, in the *dimensionless NSF atmospheric starting equations* (8.5)–(8.8), written for the velocity vector **u** and atmospheric perturbations $\pi$, $\omega$, and $\theta$, we have the balance boundary condition for $\theta$ on $z = 0$, which is (8.9) on p. 144, *Chap. 8 of our 1990 book* [22], cited in "Introduction", *entitled Asymptotic Modelling of Atmospheric Flows,* and *in the first equation (8.5) for* **u**, we have a term *proportional to* $\theta$, *viz.*,

$$- (1+\omega)\,(\mathrm{Bo}/\varepsilon)\left[1/\gamma(\mathrm{M_{st}})^2\right]\theta\mathbf{k} \equiv - (1+\omega)\left[1/(\mathrm{Fr_L})^2\right]\theta\mathbf{k}. \tag{9.8a}$$

But,

$$\begin{aligned}(\mathrm{M_{st}})^2 &= (\mathrm{U}^\circ)^2/\gamma \mathrm{R}\mathrm{T_{st}}(0),\\ (\mathrm{Fr_{L0}})^2 &\equiv \varepsilon\left[\gamma(\mathrm{M_{st}})^2/\mathrm{Bo}\right],\\ \mathrm{Fr_{L0}} &\to 0,\end{aligned} \tag{9.8b}$$

and one must have $\theta \approx (\mathrm{Fr_{L0}})^2$ in equation (8.7) of [22], cited in "Introduction", when we consider the following limiting process:

$$\mathrm{Fr_{L0}} \to 0 \quad \text{and} \quad \nu_0 \to \infty, \mathrm{Bo} \approx 1 \quad \text{and} \quad \varepsilon \approx 1, \tag{9.8c}$$

and

$$(\mathrm{Fr_{L0}})^2/(1/\nu_0) \equiv \gamma(\mathrm{m_0})^2 = \mathrm{O}(1),\ (\mathrm{m_0})^2 = (\mathrm{U}^\circ)^2/\gamma \mathrm{R}\Delta\mathrm{T_0}. \tag{9.8d}$$

The fact that $\theta \approx (\mathrm{Fr_{L0}})^2$ in equation (8.7) for $\theta$ in [22], cited in "Introduction" implies that the dimensionless stratification term is

$$\left\{(1+\pi)\left(\alpha_0/(\mathrm{Fr_{L0}})^2\right)[\mathrm{N_{st}}(\mathrm{z_{st}})]^2(\mathbf{u}\cdot\mathbf{k})\right\}, \tag{9.8e}$$

where $\alpha_0$ is a *measure for this stratification term such that*

$$\alpha_0[\mathrm{N_{st}}(\mathrm{z_{st}})]^2 \approx [\mathrm{Bo}/\mathrm{T_{st}}(\mathrm{z_{st}})]\{[(\gamma-1)/\gamma] + \mathrm{dT_{st}}(\mathrm{z_{st}})/\mathrm{dz_{st}}\}. \tag{9.8f}$$

As a consequence, with (9.8e), *one must* assume that

$$\alpha_0/(\mathrm{Fr_{L0}})^2 = \mathrm{O}(1),\ \text{assuming}\ \alpha_0 << 1, \tag{9.9a}$$

and *we can write* for the following equation for $\mathrm{dT_{st}}(\mathrm{z_{st}})/\mathrm{dz_{st}}$, according to (9.8f):

$$\mathrm{dT_{st}}(\mathrm{z_{st}})/\mathrm{dz_{st}} = -\left[(\gamma-1)/\gamma + \lambda_0\Gamma_{\mathrm{st}}(\mathrm{z_{st}})(\mathrm{Fr_{Hst}})^2,\right. \tag{9.9b}$$

where $\lambda_0 = \text{cont.}$, and $\Gamma_{\mathrm{st}}(\mathrm{z_{st}})$ is an *arbitrary function of order unity*, which takes into account a "*weak stratification*", with standard altitude $\mathrm{z_{st}} = \mathrm{Bo}\,\mathrm{z}$, while $\mathrm{Fr_{Hst}} = \mathrm{U}^\circ/(\mathrm{H_{st}}g)^{1/2}$.

The above relation (9.9b) is a *necessary constraint* on $\mathrm{dT_{st}}(\mathrm{z_{st}})/\mathrm{dz_{st}}$ if we want to derive a *consistent companion energy equation* for the leading-order term $\theta_d$ in the expansion (9.10a), when $\mathrm{Fr_{Hst}} \to 0$:

$$\mathbf{u} = \mathbf{u}_d + \ldots; \mathrm{p} = (\mathrm{Fr}_{\mathrm{Hst}})^2 \pi_d + \ldots;$$
$$\omega = (\mathrm{Fr}_{\mathrm{Hst}})^2 \omega_d + \ldots; \tag{9.10a}$$
$$\theta = (\mathrm{Fr}_{\mathrm{Hst}})^2 \theta_d + \ldots;$$

where $(\mathrm{Fr}_{\mathrm{Hst}})^2 \approx (\mathrm{Fr}_{\mathrm{LO}})^2$, because (9.7).

With the above expansion (9.10a), we obtain to *leading order the following simplified reduced working dynamical equation* (with $\mathrm{D}/\mathrm{Dt} = \partial/\partial \mathrm{t} + \mathbf{u}_\mathrm{d} \cdot \nabla$):

$$\mathrm{St}\, \mathrm{D}\mathbf{u}_d/\mathrm{Dt} + (1/\mathrm{Bo})\, \mathrm{T}_{st} \nabla \pi_d \text{ - } \theta_d \mathbf{k} = (1/\rho_{st})\, (1/\mathrm{Re})\{\partial^2 \mathbf{u}_d/\partial \mathrm{z}^2 + \Delta_2 \mathbf{u}_d + (\mathrm{Bo}/3\gamma)\, [(\mathbf{u}_d.\mathbf{k})/\mathrm{T}_{st}]\}, \tag{9.10b}$$

where $\Delta_2$ is the 2D (x, y) Laplacian, and

$$\nabla \cdot \mathbf{u}_d = (\mathrm{Bo}/\gamma)[(\mathbf{u}_d \cdot \mathbf{k})/\mathrm{T}_{st}], \tag{9.10c}$$

and where, in the troposphere, thanks to (9.9b), when

$$\mathrm{T}_{st}(\mathrm{Boz}) = 1 - [\mathrm{Bo}(\gamma - 1)/\gamma]\mathrm{z} \tag{9.10d}$$

and

$$\rho_{st}(\mathrm{z}) = [\mathrm{T}_{st}(\mathrm{Bo\ z})]^{1/(\gamma - 1)}, \tag{9.10e}$$
$$\mathrm{dlog}\rho_{st}(\mathrm{Bo\ z})/\mathrm{dz} = \text{ - } (\mathrm{Bo}/\gamma)\ [1/\mathrm{T}_{st}(\mathrm{Boz})]. \tag{9.10f}$$

Finally, to leading order, we obtain the following dimensionless energy equation for $\theta_d$, *and it seems to me that it is only by using the RAM approach that such a complicated, but significant reduced leading-order working model equation for* $\theta_d$ *can be derived*:

$$\mathrm{St}\, \mathrm{D}\theta_d/\mathrm{Dt} \text{ - } [(\gamma - 1)/\gamma]\mathrm{St}\, \mathrm{D}\pi_d/\mathrm{Dt} + \lambda_0 \mathrm{Bo}\{\Gamma_{st}(\mathrm{Bo\ z})/\mathrm{T}_{st}(\mathrm{Boz})\}(\mathbf{u}_d \cdot \mathbf{k})$$
$$= (1/\rho_{st}(\mathrm{Bo\ z}))[1/\mathrm{Pr\ Re}]\{\Delta_2 \theta_d + [\partial^2 \theta_d/\partial \mathrm{z}^2$$
$$-(2(\gamma - 1)/\gamma)][(\mathrm{Bo}/\mathrm{T}_{st}(\mathrm{Boz}))\partial \theta_d/\partial \mathrm{z}]$$
$$+\left[[(\gamma - 1)/\gamma]/(\rho_{st}(\mathrm{Bo\ z})\mathrm{T}_{st}(\mathrm{Boz}))\right](\mathrm{Bo}/\mathrm{Re})\Phi_d, \tag{9.10g}$$

where $\Phi_d$ is the limiting value of the *viscous dissipation*, and the perturbation term $\omega_d$, for the density ρ, is given by the relation

$$\omega_d = \pi_d - \theta_d, \tag{9.10h}$$

assuming that dry atmospheric air is a thermally perfect gas.

For (9.10g) as thermal boundary condition on z = 0, we write

$$\theta_d = \left[\mathrm{Bo}/\gamma(\mathrm{m}_0)^2\right]\Theta(\mathrm{t},\ \mathrm{x},\ \mathrm{y})\ \text{on}\ \mathrm{z} = 0. \tag{9.10i}$$

Note that it is easy to prove from above system of (anelastic) deep equations (9.10b, c–h,f) with (9.10d,e,f,i) that, *when Bo* $\to 0$, *we can derive the shallow Boussinesq* equations, which are *valid* and *significant only in the vicinity of* $z = 0$ *(the ground)* in a *lower atmospheric layer with thickness* $\mathrm{H}_{B*} \approx 10^3\mathrm{m}$. *This unfortunate fact about the Boussinesq approximation is a* direct consequence of the similarity relation $\mathrm{Bo}/\mathrm{M}_{\mathrm{st}} = B* = \mathrm{O}(1)$, where the atmospheric Mach number is $\mathrm{M}_{\mathrm{st}} = \mathrm{U}°/[\gamma \mathrm{RT}_{\mathrm{st}}(0)]^{1/2}$, and in such a case we have

$$\mathrm{H}_{B*} = B*[\mathrm{U}°/\gamma]\,[\mathrm{RT}_{\mathrm{st}}(0)/\gamma]^{1/2}. \tag{9.10j}$$

One the other hand, our anelastic *deep equations* derived above, with $\mathrm{Bo} = \mathrm{O}(1)$, are typically "*tropospheric equations*", valid in an atmospheric layer $\mathrm{RT}_{\mathrm{st}}(0)/g \approx 10^4\mathrm{m}$, and include the *shallow Boussinesq equations*!

*Otherwise stated, the Boussinesq equations can be "completed" by the deep equations—the latter then contain the former, which are valid only near the ground where the thermal condition* (9.10i) *holds. The curious reader can compare our anelastic (deep) equations derived above with these many very "naïve" anelastic equations which are reduced in an ad hoc manner, and for this one only has to do a search for "anelastic equations" on Google!*

## 9.5 Small-Mach-Number Time-Dependent NSF Models

Below we consider the *small* $M << 1$ limiting process, viz.,

$$\mathrm{M} \to 0\ \text{with t and}\ \mathbf{x}\ \text{fixed}, \tag{9.11a}$$

and assume that

$$\mathrm{p} = \mathrm{p}_0(\mathrm{t}) + \mathrm{M}\,\mathrm{p}_1(\mathrm{t}) + \mathrm{M}^2\mathrm{p}_2(\mathrm{t}) + \dots, \tag{9.11b}$$

where $\mathrm{p}_0(\mathrm{t}) = \rho_0(\mathrm{t})\ \mathrm{T}_0(\mathrm{t})$,
and

$$[\mathbf{u},\ \rho, \mathrm{T}](\mathrm{t}, \mathbf{x},\ \mathrm{M}) \to [\mathbf{u}_0,\ \rho_0, \mathrm{T}_0](\mathrm{t}, \mathbf{x}),\quad \text{when}\quad M \to 0\quad \text{with } t \text{ and } \mathbf{x} \text{ fixed}. \tag{9.11c}$$

In this case, instead of the full time-dependent (but with Strouhal number = 1) NSF system of equations, we obtain the following *small-Mach-number dissipative time-dependent* system:

$$\begin{aligned}&\mathrm{D}\rho_0/\mathrm{Dt} + \rho_0 \nabla \cdot \mathbf{u}_0 = 0,\\ &\rho_0 \mathrm{D}\mathbf{u}_0/\mathrm{Dt} + (1/\gamma)\nabla \mathrm{p}_2 = (1/\mathrm{Re})\,[\Delta \mathbf{u}_0 + (1/3)\nabla(\nabla \cdot \mathbf{u}_0)],\\ &\rho_0 \mathrm{DT}_0/\mathrm{Dt} + (\gamma - 1)\mathrm{p}_0(\mathrm{t})\nabla \cdot \mathbf{u}_0 = (\gamma/\mathrm{Pr\,Re})\Delta \mathrm{T}_0,\end{aligned} \tag{9.11d}$$

where $\mathrm{D/Dt} = \partial/\partial \mathrm{t} + (\mathbf{u}_0 \cdot \nabla)$.

An obvious consequence of the first and last equations of (9.11d) is the following relation for the unknown leading-order pressure $\mathrm{p}_0(\mathrm{t})$:

$$\mathrm{dp}_0(\mathrm{t})\mathrm{dt} = -\gamma \nabla [\mathrm{p}_0(\mathrm{t})\mathbf{u}_0 - (1/\mathrm{Pr\,Re})\nabla \mathrm{T}_0], \tag{9.11e}$$

and we obtain for $\nabla \cdot \mathbf{u}_0$ the equation

$$\nabla \cdot \mathbf{u}_0 = -\,(1/\gamma \mathrm{p}_0(\mathrm{t}))\,\mathrm{dp}_0(\mathrm{t})\mathrm{dt} + (1/\mathrm{Pr\,Re})\nabla(\mathrm{T}_0/\mathrm{p}_0(\mathrm{t})). \tag{9.11f}$$

Since $\mathrm{T}_0(\mathrm{t}) = \mathrm{p}_0(\mathrm{t})/\rho_0(\mathrm{t})$, instead of the third equation, for $\mathrm{T}_0(\mathrm{t})$, in (9.11d), we can use (9.11e) and derive the following system for $\mathbf{u}_0$, $\mathrm{p}_2$, and $\rho_0(\mathrm{t})$:

$$\begin{aligned}&\mathrm{D}\rho_0/\mathrm{Dt} + \rho_0 \nabla \cdot \mathbf{u}_0 = 0,\\ &\rho_0 \mathrm{D}\mathbf{u}_0/\mathrm{Dt} + (1/\gamma)\nabla \mathrm{p}_2 = (1/\mathrm{Re})\,[\Delta \mathbf{u}_0 + (1/3)\nabla(\nabla \cdot \mathbf{u}_0)],\\ &\nabla \cdot \mathbf{u}_0 = -\,(1/\gamma \mathrm{p}_0(\mathrm{t}))\,\mathrm{dp}_0(\mathrm{t})\mathrm{dt} + (1/\mathrm{Pr\,Re})\nabla(\mathrm{T}_0/\mathrm{p}_0(\mathrm{t})),\end{aligned} \tag{9.11g}$$

where $\mathrm{p}_0(\mathrm{t})$ is also present. Then, by integrating (9.11e) over a time-dependent bounded domain $\Omega(\mathrm{t})$, we obtain

$$\mathrm{dp}_0(\mathrm{t})\mathrm{dt} = -\,[\gamma/\mathrm{V}(\mathrm{t})] \iint_{\Sigma(\mathrm{t})} [\mathrm{p}_0(\mathrm{t})\mathbf{u}_0 - (1/\mathrm{Pr\,Re})\nabla \mathrm{T}_0] \cdot \mathbf{n}\,\mathrm{d}\Sigma, \tag{9.11h}$$

where $\Sigma(\mathrm{t}) = \partial \Omega(\mathrm{t})$ and $\mathrm{V}(\mathrm{t})$ is the reduced dimensionless volume of the time-dependent bounded domain (in accordance with $\rho_0(\mathrm{t}) = 1/\mathrm{V}(\mathrm{t})$).

We have thus derived a closed system of unsteady equations (9.11f) and (9.11g), of integro-differential type in the general case, for the set of flow variables ($\mathbf{u}_0$, $\mathrm{p}_2$, $\rho_0$, $\mathrm{T}_0$, $\mathrm{p}_0(\mathrm{t})$) in the bounded/closed time-dependent domain $\Omega(\mathrm{t})$. According to *Fedorchenko (1997),* one can check that this model is characterized by the *complete exclusion of any acoustic effect*, as in the case of its inviscid version. A local analysis of equations (9.11f) and (9.11g) also excludes characteristics responsible for the propagation of sound disturbances!

Indeed, we now have an incompletely parabolic system, where the second spatial derivatives cannot lead to the existence of sound waves, and a distinctive feature of the above small-Mach-number dissipative system is that $\nabla \cdot \mathbf{u}_0 \neq 0$, even if $\mathrm{dp}_0(\mathrm{t})\mathrm{dt} = 0$, and this reflects the conformity of the above model to the natural fact that any change in the temperature field due to heat conduction must be accompanied by flow. Concerning the system of equations (9.11d), it seems (see *Embid 1987*) that this system is well-posed for a judicious choice of initial and boundary conditions.

*Finally, we observe that every solution of a nonlinear problem in fluid dynamics, especially by the time-dependent solution of the NSF equations, has a high intrinsic value, irrespective of its practical interpretation!* However, in *combustion problems*, one must take into account *fast acoustic waves*, which emerge in the initial stage of unsteady flow, and their influence on the *averaged quasi-incompressible* flow. In the dissipative case, this is a *very difficult* problem, and we are actually *only able to derive a full consistent model for the non-viscous case (see Sect. 5.3* in our LNP 672, 2006, [24], cited in "Introduction").

## 9.6 Analysis of Through-Flow Equations

In 1974 <1>, we take into account the *analogy between the equations of the mean flow in an axial turbomachine, viz.,* (9.12a)*, that is, the symmetrical through-flow equations obtained by a suitably defined averaging operation over the azimuthal variation, assuming a very large number of blades in the row, and the equations governing thermally nonconducting flows of a compressible ideal (Eulerian) fluid. To begin with, such an analogy allows one to explain the mechanism initiating the vortex (vorticity) inside the row, and it picks out characteristics pertaining to the vortex field. It also allows one to obtain a general representation of the velocity field of the mean flow in the row, identically satisfying the equation of motion of this flow.*

*Starting from work by Hollmann <2>, one writes down the independent system of Lagrangian invariants corresponding to the mean flow equations, describing the evolution of five scalar fields along the fluid trajectories of the mean, unsteady, symmetrical through-flow. Finally, one considers the system of first integrals corresponding to the case of steady mean flow!*

According to the results of Sect. 8.2, with an *error of* $\varepsilon^2$, from the *axially symmetrical through-flow equations* (8.9b,c), with equation (8.9h), we have

$$\begin{aligned} &\partial \Delta/\partial t + \mathrm{div}(\Delta < \mathbf{v} >) = 0;\\ &\partial < \mathbf{v} > /\partial t + [\mathbf{Curl} < \mathbf{v} > + 2\Omega \mathbf{e}_z] \wedge < \mathbf{v} > + \mathbf{grad} J\\ &- (1/\Delta)\Pi\, \mathbf{grad}\, \Sigma = 0;\\ &\partial \Sigma/\partial t + < \mathbf{v} > \mathbf{grad}\, \Sigma = 0. \end{aligned} \tag{9.12a}$$

We observe that the above system of three equations (9.12a) is very similar to the system of equations governing a non-viscous compressible, but non-heat-conducting fluid, if we introduce the pseudo-*density* $\rho$, *pseudo-temperature* T, and pseudo-*specific entropy* $S$ such that

$$\rho = \Delta, \;\; \Pi/\Delta = \mathrm{T} \text{ and } S = \Sigma. \tag{9.12b}$$

The functions Δ and Σ are given functions related to the geometry of the row of the blades, with $\Pi \neq 0$. If

$$\boldsymbol{\omega} \equiv \mathbf{Curl} < \mathbf{v} > +2\Omega \mathbf{e}_z, \tag{9.12c}$$

then instead of the second equation in (9.12a), we can write

$$\begin{aligned} \mathrm{D}(\omega/\Delta)/\mathrm{Dt} = [(\boldsymbol{\omega}/\Delta)\cdot \mathbf{grad}] < \mathbf{v} > \\ + (1/\Delta)\,[\mathbf{grad}\,(\Pi/\Delta) \wedge \mathbf{grad}\,\Sigma]. \end{aligned} \tag{9.13a}$$

This last equation (9.13a) can be used to analyse the mechanism whereby the vorticity $\boldsymbol{\omega}$ emerges inside the row in through-flow: vorticity is created as often as each particle crosses a region where the surfaces $\Pi/\Delta = \text{const.}$ and $\Sigma = \text{const.}$ intersect! More precisely, one can write

$$\mathbf{grad} \wedge [(\Pi/\Delta)\,\mathbf{grad}\,\Sigma] = \mathbf{grad} \wedge \boldsymbol{F}, \tag{9.13b}$$

where $\boldsymbol{F}$ is given by (8.14a), with (8.14e) (see Sect. 8.2). If we now introduce the function λ defined by

$$\mathrm{D}\lambda/\mathrm{Dt} = \Pi/\Delta, \tag{9.13c}$$

and also the vector

$$\mathbf{U} =< \mathbf{v} > \text{-}\lambda\,\mathbf{grad}\,\Sigma + \Omega\,\mathrm{r}\,\mathbf{e}_\theta, \tag{9.13d}$$

then we can rewrite the second equation for <**v**> in (9.12a) in the following form:

$$\partial\mathbf{U}/\partial\mathrm{t} + (\mathbf{grad} \wedge \mathbf{U}) \wedge \mathbf{v} = \mathbf{grad}\,\chi, \tag{9.13e}$$

with

$$\mathbf{U} = \mathbf{grad}\,\phi \textit{ and } \partial\phi/\partial\mathrm{t} = \chi \tag{9.13f}$$

For <**v**>, we obtain

$$< \mathbf{v} >= \mathbf{grad}\,\phi + \lambda\,\mathbf{grad}\,\Sigma - \Omega\,\mathrm{r}\,\mathbf{e}_\theta, \tag{9.14a}$$

an equation which gives the structure of <**v**> inside the row (where $\lambda \neq 0$). The scalar functions ϕ, λ, and Σ in (9.14a) satisfy

$$\nabla^2\phi + \lambda\nabla^2\Sigma + \nabla\Sigma\cdot\nabla\lambda + (1/\Delta)\,\mathrm{D}\Delta/\mathrm{Dt} = 0, \tag{9.14b}$$

as a consequence of the first equation for Δ in (9.12a). Outside of the row, when $\lambda = 0$ and $\Delta = 1$, we recover the Laplace equation $\nabla^2\phi = 0$

In the *unsteady case* considered above, according to the third equation in (9.12a), *the function $\Sigma$ is a Lagrangian invariant*, i.e., $D\Sigma/Dt = 0$. Concerning the theory of Lagrangian invariants, see *Chap. II of our* 1974 *book* [8], cited in Chap. 1.

For the present case, we can use $\Sigma$, and obtain a second *Lagrangian invariant:*

$$D[(\boldsymbol{\omega}/\Delta)\cdot\nabla\Sigma]/Dt = 0 \tag{9.14c}$$

and in <1>, the reader can find a representation of our *axially symmetrical through-flow system of equations* (9.12a) with five *Lagrangian invariants.*

These three complementary *Lagrangian invariants are derived from the Ertel generalized continuity equation in 1960 <3>.*

In the *steady case* ($\partial/\partial t = 0$), from the system of equations (9.12a), when both $\partial\Delta/\partial t = 0$ and $\partial\Sigma/\partial t = 0$, in the first and third equations, we can write

$$<\mathbf{v}> = (1/\Delta)[\nabla\Psi \wedge \nabla\Sigma] \tag{9.15a}$$

where $\Psi$ is the stream function for the *axially symmetrical through-flow and we have the relations*

$$u = (1/r\Delta)\partial\Psi/\partial z \text{ and } w = -(1/r\Delta)\partial\Psi/\partial r, \tag{9.15b}$$

Instead of the second equation for $\langle\mathbf{v}\rangle$ in (9.12a), we obtain the two first integrals

$$\boldsymbol{\omega}\cdot\nabla\Psi = -\Pi \text{ and } \boldsymbol{\omega}\cdot\nabla\Sigma = -\Delta\, dJ/d\Psi. \tag{9.16}$$

From (9.16), one can obtain an equation for $\Psi(r, z)$ in the form

$$\begin{aligned} &\left[1 + (r\partial\Sigma/\partial r)^2\right]\partial^2\Psi/\partial z^2 - 2\,r^2(\partial\Sigma/\partial r)\,(\partial\Sigma/\partial z)\partial^2\Psi/\partial z\partial r \\ &+ \left[1 + (r\partial\Sigma/\partial z)^2\right]\partial^2\Psi/\partial r^2 + D(r,z)\,\partial\Psi/\partial z + E(r,z)\partial\Psi/\partial r \\ &= 2r^2\Delta\,\omega\,\partial\Sigma/\partial z - (r\Delta)^2 dJ/d\Psi, \end{aligned} \tag{9.17}$$

where the coefficients D and E are known. In fact, equation (9.17) is similar to an equation due to Sirotkin (1966), derived in a rather (ad hoc) manner, and also the equation obtained in ONERA by R. Legendre in another way!

## References


<1> Zeytounian RKh (1974) Analyse des equations régissant l'écoulement moyen dans une roue de turbomachine axiale. Bull Acad Polonaise Sci Sér des Sciences Techniques XXII (4) [see also pages, 316–322 in LNP 27(1974)]

<2> Hollmann G (1964) Arch Meteor Geophys Bioklim A14:1–13

<3> Ertel H (1960) Gerl Beitr Geophys 69: 357–361

# Epilogue

Although one of the main motivations of this book is the *consistent and rational modeling* of very stiff fluid dynamics problems, to provide meaningful and valuable numerical computation methods for Numericians, one should not overlook the role of the RAM approach in the *fundamental problem* of unifying and classifying the family of standard *Newtonian fluid dynamics models* via the *deconstruction* of the *Navier-Stokes-Fourier* (NSF) system of equations for a compressible, viscous, and heat-conducting fluid!

The main governing idea in the above chapters was that the various working models of *classical fluid dynamics* must be obtained as significant degeneracies of the Navier-Stokes-Fourier (NSF) system of equations. Hence, these equations play a fundamental central role here and also in our two companion books, published in 2012 and 2014 by Springer-Verlag. In fact, this role of the NSF equations is not only fundamental, but truly central!

The *main objective* was to "*deconstruct*" this *Navier-Stokes-Fourier system* in order to *unify and classify* the various partial simplified approximate models used in traditional Newtonian classical fluid dynamics, and this *first facet* is clearly a challenge that will have a very important *pedagogical impact* on university education.

The *second facet*, involving by the "*deconstruction*" *of the NSF system,* is to outline a (rather elementary) rational consistent asymptotic theory of the *origin* and also *classification* of working models for various fluid flows, using the emerging family of non-dimensional parameters in various dimensionless initial-boundary value NSF problems.

The *third facet* is illustration of the RAM approach as a mathematical tool in the derivation of various stiff technological and geophysical problems in aerodynamics, thermal and thermocapillary convections, nonlinear acoustics, and meteo-fluid dynamics, from the corresponding NSF problems.

This can in fact be considered as an aid to Numericians confronted with the very difficult numerical computations in super-high-power computers! It is important to establish the consistency of the model used by numericians and its suitability as a



R.Kh. Zeytounian, *Challenges in Fluid Dynamics*,
DOI 10.1007/978-3-319-31619-2

valid substitute for reality, as represented by the full NSF system. Our purpose in the present book was to initiate a process which does not seem to have attracted sufficient attention from scientists working on fluid dynamics modeling of very stiff initial and boundary value problems.

A paradoxical aspect of fluid dynamics as a branch of mathematical physics is the widespread ignorance of the usefulness of the NSF equations governing the unsteady motion of a compressible, viscous, and heat-conducting Newtonian fluid. From Chap. 1, the reader is now well informed about the "*adventure*" of the discovery of this NSF system, from Newton, to Euler, Cauchy, and Navier, then Stokes and Fourier!

Today, despite tremendous progress with initial and boundary value NSF problems, and 82 years after Leray's two pioneering fundamental papers of 1933 and 1934, progress on the purely rigorous mathematical aspects turns out to be much less helpful on the practical level!

RAM is a *word* I came up with in the late 1970s to describe a rational methodology that could extract unsteady consistent reduced/simplified working fluid flow models from the NSF equations governing a compressible, viscous, and heat-conducting fluid. In fact, it turned out that the RAM approach could also answer the question raised by Shinbrot in 1973 (see the references for the preface).

The methodology in question is based on three essential concepts: *models*, *asymptotics*, and *rationality*. By models, I understand *first* and foremost an NSF problem, posed in a given region of space-time and specified by a system of *partial differential equations*, with *boundary conditions and initial data. Second, I understand* the above-mentioned reduced/simplified working fluid flow models. *Asymptotics is the main tool for obtaining these* reduced/simplified working fluid flow models from NSF, and the question of *rationality* refers to the *non-ad hoc manner in which these* reduced working fluid flow models are derived.

An NSF problem is supposed to have a definite solution, not in the framework of rigorous mathematical analysis, but through experience of reduced working fluid flow model equations with associated relevant initial and boundary conditions as provided by observation, calculation, and simulation, and especially by numerical computation using *ultra-powerful* computers with a very *large memory*. It is clear that this experience comes from a long activity of research in fluid dynamics and university teaching. The starting NSF problem is considered in Chap. 1. It governs Newtonian fluids satisfying the concept of "fluidity" introduced by Stokes and Fourier's law as regards the heat vector.

In fact, according to the RAM approach, many other, simpler reduced models result from the NSF system. They are simpler in the sense that their solution is more readily accessible to numerical calculation, and these model systems are indeed very significant degeneracies of the NSF equations. In truth, with the RAM approach, one must first specify a systematic methodology, described here as rational, to infer various *less stiff* reduced models for various fluid flows from the original NSF system, each applicable in a limited region of the domain where the starting NSF problem is specified!

However, we have no mathematical method for characterizing the various regions where the specific models emerging from NSF apply. Past experience plays a fundamental role, but it seems difficult to provide guidance concerning this characterization. On the other hand, the regions in question usually have a simple shape and can be characterized by three lengths and a time. In meteorology, for example, we must introduce the curvature of space, thereby including the radius of the earth as a length and introducing the parameter called "*sphericity*", and also, of course, gravity (related to the Froude and Boussinesq numbers) and the Coriolis force related to the Rossby number.

Regarding the time, the global domain on which the problem is posed is broken down into sub-regions, involving an assumption that also has no mathematical basis. The hypothesis is to admit that each of the variables and their various derivatives appearing in the NSF equations have both positive and negative values in each sub-domain, these being neither close to zero nor close to infinity. But since the values in question depend on the choice of the units, one must first obtain dimensionless equations and dimensionless initial data and boundary conditions for the overall NSF problem.

This is where we introduce the *dimensionless* problem with certain non-dimensional parameters, and at the same time, reveal the rationality of our RAM approach, taking into account various limiting processes relating to various non-dimensional parameters, assumed small or large relative to unity. Unfortunately, this approach also leads to various singularities which divide the domain of existence of the NSF problem into sub-domains specific to reduced/simplified working models. The problem of initial conditions and specific boundary conditions for each reduced system of equations in its associated sub-domain remains a priori a difficult problem, and is based mainly on acquired knowledge.

Naturally, every reduced problem associated with its sub-domain must have a unique solution, at least for a finite time interval. It is at this point that the mathematician can help the numerical analyst, and interdisciplinarity brings together the fluid dynamics theorist, mathematician, and numerical analyst!

As mentioned above, one great advantage of the RAM approach relates to the possibility of generalizing the various solutions of interesting well known incompressible or non-viscous problems to slightly compressible or weakly viscous ones. To this end, starting from the corresponding full NSF formulation, the RAM approach provides a way to derive an associated consistent second-order model problem alongside the leading-order model problem. Here, certain terms take into account the effects of the compressibility or the viscosity, neglected (as small) in the leading-order model problem.

To conclude, I hope that the motivated and curious reader will have understood the following points:

*The main governing idea of this book Challenges in Fluid Dynamics is that, on the one hand, the many and varied existing teachable models of flows, both used and useful, must be obtained rationally as significant degeneracies of well-posed NSF problems, and on the other hand, we have in our hands a very effective tool for the modeling stiff technological and geophysical flows via numerical simulation with the help of super-high-speed computing.*

*Clearly, only the derivation of a consistent, significantly reduced fluid dynamics model problem, emerging from NSF, for a physically real stiff problem, can guarantee that we will obtain a "good" numerical simulation!*

Fourteen years ago, I wrote (see page XV of my book published in FMIA 64): "*I am strongly convinced that, over time, the growth in capabilities of numerical simulation will depend on, or at least be related to, the development of the RAM approach.*" And indeed the publication of my books in 2006, 2009, 2012, and 2014 demonstrate the validity of this statement!

Finally, my *presentation* of fluid dynamics in this book, where the *main new argument relates to the central role of the NSF initial-boundary value problem*—thus providing a way to deeply understand the *origin* of traditional and useful models of Newtonian classical fluid dynamics and their classification—is in fact an *answer* to a question I raised *sixty years ago*, in Yerevan (Armenia), where I sought to understand the *real role* of the NSF equations and wondered *why* this fundamental system of equations was not being investigated, in a practical way, in all books devoted to theoretical fluid dynamics!

Our sincere hope is that this *Challenges in Fluid Dynamics* will be *helpful* to *motivated* and *curious students, as well as young scientists interested in the analysis of various fluid flows. I am also sure that numericians confronted with the computation of very stiff fluid flow problems will find much useful information here.*

Concerning *pure mathematicians*, and in particular *young mathematicians, who want to prove rigorous theorems relating to the existence and uniqueness of fluid dynamics problems, it is obvious (at least to me) that they will find here many relevant fluid dynamics problems. Such an approach certainly has a practical interest when coupled with numerical computation of these models as considered by mathematicians!*

This Challenges in Fluid Dynamics is probably [aged: Born in 1928 (Paris—France) from Armenians parents] the last book! I hope that the reading of this challenges will tempt to read my preceding other books published during the years 1974's to 2012 and especially my Scientific Autobiography edited in 2014 by Springer at Heidelberg.

# Index




R.Kh. Zeytounian, *Challenges in Fluid Dynamics*,
DOI 10.1007/978-3-319-31619-2

The manufacturer's authorised representative in the EU is Springer Nature Customer Service Centre GmbH, Europaplatz 3, 69115 Heidelberg, Germany. If you have any concerns regarding our products, please contact ProductSafety@springernature.com

Printed and bound by CPI Group (UK) Ltd, Croydon, CR0 4YY
25/09/2026
02222854-0018